The Miegunyah Press
at
Melbourne University Press

The General Series of the Miegunyah
Volumes was made possible by the
Miegunyah Fund established by
bequests under the wills of
Sir Russell and Lady Grimwade

'Miegunyah' was the home
of Mab and Russell Grimwade
from 1911 to 1955

Backyard Insects

PAUL A. HORNE
DENIS J. CRAWFORD

THE MIEGUNYAH PRESS

AT

MELBOURNE UNIVERSITY PRESS

Melbourne University Press
PO Box 278, Carlton South, Victoria 3053, Australia

First published 1996

Designed by Lauren Statham, Alice Graphics
Typeset in 10.5/16.5 pt ITC Giovanni Book
Printed in Australia by McPherson's Printing Group

National Library of Australia Cataloguing-in-Publication entry

Horne, Paul A. (Paul Anthony), 1956– .
   Backyard insects.
   Bibliography.
   Includes index.
   ISBN 0 522 84737 4.
   1. Insects. 2. Insects—Identification. I. Crawford,
   Denis J. (Denis Julian), 1957– . II. Title.
595.7

CONTENTS

ABOUT THE PHOTOGRAPHS

All photographs were taken by Denis Crawford with a few exceptions. The photographs on pp. 90 and 158 were supplied by the Australian National Insect Collection, CSIRO Division of Entomology (© Magnus Peterson CSIRO and ANIC Photo Library, respectively). Other photographs were taken by Colin Bower (p. 190), David Paul, Department of Zoology, University of Melbourne (p. 240) and Gary Yu (pp. 216 and 218).

A few photographs appear without identification. The illustrated insects are: a common imperial blue butterfly (opposite the title page); a European wasp (p. viii); a katydid (p. 6); a greengrocer cicada (p. 40); a Christmas beetle (p. 68); a robber fly (p. 108); the head of an adult emperor gum moth (p. 134); an emperor gum moth caterpillar (p. 136); and a bull ant (p. 180).

ACKNOWLEDGEMENTS

This book was only possible with the support and assistance given by many people, particularly those who helped with the collecting of insects. We would like to especially thank: J. Bentley, V. Carter, P. Cole, J. and K. Comery, E. J., M. and P. Crawford, N. Dowsett (butterfly keeper, Melbourne Zoo), C. and J. Edward, Dr W. Frost, D. Glenn, Janet and James Horne, A. and S. C. Horne, S. Keogh (Melbourne University Press), S. Kreidl, Dr M. Malipatil, C. Mansfield, S. Narain, Dr I. D. Naumann, Dr T. R. New, D. Papacek, Dr A. M. Smith, G. Smith, S. Tymms, C. Symington and Dr J. R. Woodward.

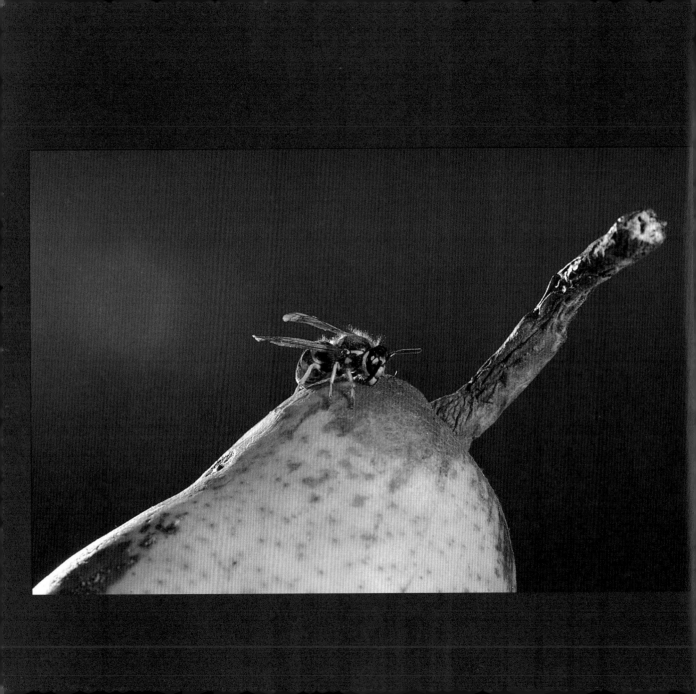

INTRODUCTION

IN EVERY HOUSE AND GARDEN in Australia there are many species of insects. A few, such as clothes moths, blowflies, mosquitoes and aphids, are unwanted guests because of the damage they cause or the nuisance they create. As these are the insects that we may have most contact with, the impression that many people have is that almost all insects are pests. This is definitely not the case, as the number of pests are vastly outnumbered by species that are beneficial or which have no obvious direct impact on our lives. Some insects are directly helpful, for example by killing pests or pollinating plants, while others are indirectly helpful by contributing to other processes (for example, as food for others, or by decomposing or recycling plant material).

There are many types of insects but how do you recognize them? Some insect specialists (taxonomists) describe and define what makes one species different from another. Closely related species are sometimes easily distinguished from one another, but more commonly are very often only distinguishable by minute differences that can be seen only with microscopes and a trained eye. This book is not written for the specialist but for anyone who wants to know a little more about the insects that share our homes and gardens.

Many books about insects begin by stating how numerous and diverse they are, because this is one of their most striking aspects. How many insects are there in Australia? One recent estimate is more than 85 000 species. This compares to about 250 mammals and over 750 species of birds. Even if a small fraction of all Australian insects exists in and around one city it would be an enormous task just to catalogue the species we find in its gardens.

The intention of this book is to allow anyone to recognize some different insects that commonly occur in the major urban areas of Australia and at the same time to show some of the interesting aspects of insects. We have chosen some species as examples of the larger groups (orders and families) of insects so that you will be able to see what makes one group different from another.

To achieve this we have prepared full colour photographs of one hundred species (ninety-four insects plus a few non-insect species). It should allow you to compare the insect in your backyard with the one in the book. If it is not exactly like the one in the book, but similar, then it probably belongs to the same group. The book will help you to find out a little more about your immediate environment, and serve as an introduction to insects for those who want to learn more. We have tried to produce a practical handbook that can be used easily by people in most urban areas of Australia.

The first sections of the book contain general information about insects and at the end of the book is a list of further reading and contacts for those wanting to find out more about entomology (the study of insects). There is also a glossary. We have deliberately tried to keep the use of specialist terms to a minimum, but a few are used in some descriptions. The non-insect species that we have included are species that are commonly found with insects, and often confused with them, and these species serve as examples of their groups.

Most of us learn at school that insects have six legs and spiders have eight legs, but there are quite a few other types of animals that are often grouped together as 'bugs'. Along with other animals (such as snails, shrimps, millipedes and worms) both insects and spiders are invertebrates: they do not possess a backbone or spine. Insects and spiders have jointed legs and, instead of an internal backbone, they have a hard skin that also functions as a supporting skeleton (an 'exoskeleton').

The insects are divided into groups (called orders) based on their physical appearance. Entomologists (people who study insects) in Australia generally recognize twenty-nine different orders, and these could be divided into primitive-wingless orders (two), primitive-winged orders (two) and modern-winged orders (twenty-five). Insects with primitive wings cannot fold them over their body. The modern-winged insects are further divided on the basis of the way in which the immature insect develops. As with any rule, there are exceptions to confuse the situation. Some of the modern-winged insects have no wings at all (fleas and lice) and some cannot fold their wings over the body (butterflies).

Among the modern-winged insects there are fourteen orders in which the young insects resemble the adults, such as the true bugs (Hemiptera). There are another eleven orders in which the mature insects are completely different in appearance from the adults and must pass through an intermediate stage, the pupa, to complete their development.

Some insects orders are either not found in Australia or not readily encountered, and we have not included any examples of these groups. We have omitted many aquatic insects so that we could include some that are more commonly encountered in the urban environment. We have chosen what we believe are some of the more interesting and commonly found species of insects, representing eighteen orders. These are:

- silverfish (Thysanura)
- dragonflies and damselflies (Odonata)
- cockroaches (Blattodea)

- termites (Isoptera)
- praying mantids (Mantodea)
- earwigs (Dermaptera)
- crickets and grasshoppers (Orthoptera)
- stick insects (Phasmatodea)
- psocids or booklice (Psocoptera)
- lice (Phthiraptera)
- true bugs (Hemiptera)
- thrips (Thysanoptera)
- ant-lions and lacewings (Neuroptera)
- beetles (Coleoptera)
- fleas (Siphonaptera)
- flies (Diptera)
- butterflies, skippers and moths (Lepidoptera)
- bees, wasps, sawflies and ants (Hymenoptera)

We have also included some other common invertebrates that are often mistaken for insects, or thought of as insects, so that you can see the differences and recognize another few groups of animals.

We give the scientific name of the insect illustrated with its description. The name always consists of two parts, the *Genus* and then the *species*. For example, the common brown butterfly has the scientific name *Heteronympha merope*. The scientific name is used to avoid confusion about what insect is being referred to. There may well be butterflies called 'common brown' in other parts of the world but which are different species. The scientific name refers only to a single species. It is a useful way for scientists, in particular, to be sure that other scientists know precisely what species they are talking about. The common name, if one exists, is usually sufficient for people from the one area to know what insect is being discussed.

Similar species are grouped in a genus, similar genera are grouped in a 'family', and similar families make up an order. Entomologists also use a number of other intermediate catagories in this hierarchy. The common brown butterfly, *Heteronympha merope*, belongs to the genus *Heteronympha*, the family Nymphalidae and the order Lepidoptera. You can recognize a family name as, by convention, it always ends in -idae, while a genus is printed in italics beginning with a capital letter. The species name is in italics without a capital letter. Where we have chosen an insect to illustrate a family we do not use its genus and species names.

For each insect described and illustrated we have set out at the top of the page the following details:
- common name
- scientific name
- the size of the pictured insect; what it feeds on; which order it belongs to and, where appropriate, other details of its scientific classification.

THE
INSECTS

Silverfish

Ctenolepisma sp.

Entomologists classify silverfish among the most primitive, in evolutionary terms, of all insects. They have no wings at all, even the adults. Despite this interesting feature, silverfish have been relatively poorly studied, due in large part to the fact that most species run so fast when discovered! It is very difficult to study something that cannot be collected, held or watched.

So far, fewer than thirty species of silverfish have been formally identified from Australia, but most people will have found them in their cupboards or sinks. There are only a few pest species, the most common being in the genus *Ctenolepisma*. There are also several introduced species that are found all around the world.

Young silverfish look very much like small versions of the adults. After a number of moults (usually more than ten) they mature. After the first few moults they begin to develop silvery scales (which is the reason for their common name). If you pick them up, the scales will come off on your fingers (the same thing happens with moths and butterflies, whose wings are also covered in scales).

Most silverfish have three long tails which are called cerci. The common household species are known to feed on paper and glue, among other things. They are able to do this by producing an enzyme to digest cellulose, an ability which is unusual in insects.

20 mm
Generalist feeder
Thysanura

Dragonflies and damselflies
Anisoptera (dragonfly); Zygoptera (damselfly)

Brightly coloured and fast-flying dragonflies and damselflies are spectacular insects. The winged adults are often found near streams and lakes or dams where they chase and catch other insects in acrobatic displays. Dragonflies (shown in the photograph opposite) are usually larger than damselflies (shown on the next page) and may be found quite a distance from water. Their legs are arranged so that they can seize prey (such as smaller flying insects) in mid-flight. They are more delicate and slender than other insects that behave in similar ways (such as robber flies).

60 mm (dragonfly)
40 mm (damselfly)
Predators
Odonata

The wings of dragonflies and damselflies are unlike most other insects, as they cannot be folded back over the main part of the body (the abdomen) when the insect rests. Instead, these insects hold their wings vertically above their bodies (most damselflies) or flat as if they were flying (most dragonflies). The very first flying insects of prehistoric times are believed to have had similar wings, and dragonflies and damselflies are the living representatives of this primitive form. Just to confuse things, butterflies, one of the most modern and recently evolved insect groups, have redeveloped the ability to hold their wings vertically rather than folded back over the abdomen.

Immature dragonflies and damselflies are well known to anyone who has investigated the bed of a creek, and also to anglers searching for bait. Commonly known as 'mud-eyes', young dragonflies look nothing like their adult forms. They are ferocious predators, ambushing a variety of small animals under water.

To breathe underwater, young damselflies have three feathery gills on the tips of

their abdomens. The gills of dragonflies are hidden inside their bodies. Only a very few Australian species can live totally out of water. Pairs of dragonflies and damselflies are often seen together, as they mate in flight and often rest on nearby twigs or branches. Males often guard a territory and will keep other males away.

American cockroach
Periplaneta americana

One of the most common pest cockroaches that we see in our homes is an introduced species of American cockroach. It has adapted very well to this country, particularly to Sydney and the surrounding areas. However, if it is any consolation, an Australian cockroach is now a pest in America!

Cockroaches normally live in the bush but have become very used to urban environments. Many Australian cockroaches are wingless as adults, including some extremely large, burrowing species. The species we know as pests represent only a small fraction of the total number—more than 400—of species in Australia.

Many cockroaches, including the pest species, live for several years. They begin life in an egg case called an ootheca. Adult females may often be seen with a partially formed ootheca protruding from the tips of their abdomens. Each ootheca contains several eggs (usually between fifteen and twenty-five).

The ootheca are laid in sheltered places, and sometimes glued in place. Many female cockroaches will guard the newly laid, soft ootheca, most likely from other cockroaches. Young cockroaches look like smaller versions of the adults, except for the wings if adults have them. Cockroaches will eat just about anything in our houses, including, it is reported, the eyelashes and fingernails of sleeping humans.

On the positive side, although it may not be readily available now, an extract of cockroaches was not so long ago said to be useful for treating whooping cough, warts and boils (Merck's Index, 1907).

35 mm
Scavenger
Blattodea

Termite, 'white ant'
Coptotermes sp.

Insects that can eat your house are a great worry and, throughout Australia, significant amounts of money are spent on protective measures. There are over 200 species of termites in Australia, about twenty of which cause economic concern. The greatest problems are caused by species of *Mastotermes* (in Darwin and far northern Australia) and *Coptotermes* (everywhere else).

Termites are not ants, despite their other misleading common name; they are another order entirely. Termites are social insects, like bees, with different members (castes) of the same species having different roles. There are soldiers, workers, queens and kings, each with a special function in the colony. For example, a soldier of *Coptotermes* produces a defensive chemical from the front of its head when it or the colony is threatened.

A colony is founded by a winged queen and king. You will almost certainly have seen mass flights of termites (all potential kings and queens of a new colony). When the weather conditions are right—often humid, hot weather just before rain—existing colonies release thousands of winged adults at the same time. Termites must sense the change in air pressure associated with changing weather conditions.

The air is thick with weakly flying termites for a short time. This strategy means that there are more termites than predators can cope with and so some will avoid being eaten. Very soon after landing the termites shed their wings, and males and females form pairs. These pairs look for a suitable site for a new colony, but extremely few of those released from existing colonies will be successful. They usually look for damp,

8 mm
Eats wood and grasses
Isoptera

rotting wood in which to build, and so most houses are not at immediate risk. Once established in a new colony the original founding pair may live for seventeen years or more.

Termite colonies are often in mounds and are very complex structures. They have air conditioning in hot climates (by vanes on the mound) and waste disposal areas as well as royal rooms and creches. Some overseas species even grow fungus gardens. More than a million individuals may live in a single colony. Colonies may be in the obvious mounds of the tropics—this photograph shows a nest made by the spinifex termite, *Nasutitermes triodiae*—or underground, or inside or against trees.

Termites are very unusual in that they can digest cellulose—a major component of plants—without the help of protozoa (a group of microscopic organisms). Most other animals that eat wood rely on the protozoa living in their gut to produce the enzymes necessary to break down the cellulose.

From the colony, some species construct covered walkways to where they collect wood or grasses. The walkways make them hard to detect when they attack houses, and provide protection from the weather and predators.

20

Green mantid
Orthodera ministralis

Praying mantids apparently kneel with hands together, as if they were praying. They have large eyes on a triangular-shaped head, which give them a good view of potential food (usually other insects, but sometimes lizards and frogs for the larger species). Both the adults and immature insects look similar, except that the adult males usually have fully functional wings while the females and nymphs usually only have wing buds or no wings at all.

Mantids are predatory and most rely on camouflage and their ability to remain very still to catch prey. Usually they ambush or stalk their prey, grasping it with their fore legs. These fore legs have opposable segments—which work like our fingers and thumb—and are armed with rows of stiff spines to help subdue and hold their victims. Some species are territorial, and a brightly coloured spot on the inner fore leg may signal to other individuals of the same species.

Mantids are quite commonly seen in low numbers. They are general predators and will eat other members of their own species. Females are known to bite off the head of the male while mating! The nerves that control copulation are in the male's abdomen and so mating can continue unaffected.

Another unusual feature of mantid reproduction is that the female deposits her eggs in a foam. The foam hardens into a tough, protective case (an ootheca). The ootheca can be found glued to walls, trees, grass stalks or even the ground. They can often be found with many small holes in the sides, which is evidence of wasp parasites.

40 mm
Predator
Mantodea

European earwig
Forficula auricularia

The European earwig is now found almost all over the world, and is a common garden pest. It is better known to most of us than our native species because it lives close by and attacks our garden plants and fruit, and even our carpets!

European earwigs love to eat flower petals and fruit (which makes them notorious among gardeners). However, it is less well known that they eat other insects including pests such as mealybugs. Scientists have, in experiments, deliberately released these earwigs into orchards as biological control agents.

The immature earwigs (nymphs) look like the adults, and do not have a pupal stage. Like other earwigs, the adults have wings, the front pair forming covers for the neatly folded hind wings.

At the tip of their long abdomens, as in all members of their order, is a pair of forceps. Those of males are broader than those of the females. They can use these in mating (males battling for a female) but more commonly they use them to grasp food.

Females guard their eggs and young from other insect predators, including others of the same species. In warm climates European earwigs from the youngest nymphs to full-grown adults spend the hottest months sheltering in large groups (aggregations) and can easily be found under pieces of wood and tin. This practice is called aestivating (not active over summer) as opposed to the better-known practice of hibernating (not active over winter).

14 mm
Eats plants, scavenger
Dermaptera

Common brown earwig
Labidura truncata

Our experience of earwigs is most often with the introduced European earwig (see p. 23). But there are many more native species that are not pests or nuisances. One of these is the widespread species *Labidura truncata*. Its most notable feature is the orange triangle behind its head.

Labidura truncata is a fairly large species (about 30 mm long) that loves to eat soft-bodied insects like caterpillars. It seizes them with the pair of large forceps on the tip of its abdomen, and holds them there while it feeds on them. They attack and easily kill *Heliothis* caterpillars larger than themselves. If hungry enough *Labidura truncata* will eat other tougher insects, including other earwigs. For this reason, female earwigs remain with their eggs and young nymphs, to protect them from larger earwigs and other predatory insects.

Immature earwigs look much like the adults, only smaller. You can distinguish males and females by their different forceps (the males have larger forceps with one or two barbs or teeth). When fully adult, earwigs develop wings. The hind wings are folded in a neat, complex way beneath the short fore wings which are really just protective wing covers.

30 mm
Predator,
scavenger
Dermaptera

Katydid
Caedicia sp.

Not all katydids are plant feeders, but this one eats leaves. It even looks like a leaf, and relies on this camouflage for protection. In the daytime these katydids are fairly quiet and hard to find. As they are quite large insects, they are probably hiding from birds that would find them to be a very good meal. They are more easily found (by people) at lights at night. (Another photograph of this katydid appears on p. 6.)

Katydids are normally active at night, like many of their close relatives, the crickets, but unlike many grasshoppers and locusts, which also belong to this order. Katydids generally sing at night. Their songs can be very loud but can also be in frequencies beyond the range of human hearing. They produce the sound by rasping together special structures on their wings.

Like other members of this group, katydids have large hind legs for jumping. They do not tend to hop around like other crickets and grasshoppers, but either fly or walk. Katydids have very long antennae.

Female katydids may have a long tube-like apparatus on the tips of their abdomens (just as female wasps do) with which they lay their eggs. They use these to insert their eggs into an appropriate material (for example, soil or leaves).

40 mm
Eats plants
Orthoptera

Black field cricket
Teleogryllus commodus

Two species of *Teleogryllus* crickets are common in eastern Australia. They are more noticeable in summer when their normal food (grass) dries out and they start searching for new food supplies. They fly to lights on warm nights, and so are often seen where bright lights are kept on all night (for example, car showrooms, motels, street lights).

Crickets and grasshoppers are very similar, with large hind legs for jumping. Crickets have long antennae but those of grasshoppers are fairly short. The immature insects (nymphs) look much like the adults except that they don't have large wings.

Female crickets can be distinguished from males by the long ovipositor at the tips of their abdomens. They use this to deposit their eggs deep in the soil where they may lie for many months in the cooler southern areas. The eggs have a resting, dormant stage (called a diapause) to avoid the cold harsh months. Nymphs hatch in the spring, in time to eat the new flush of plant growth.

Males are responsible for cricket songs. They make the sound by rasping one wing against the other. Then they listen with the equivalent of our ear located on their fore legs! Because only males produce songs, the pattern of veins on the wings of males looks quite different from those on females. The type of song produced is unique to a species, and is sometimes used to distinguish species where no visible characteristics are useful.

30 mm
Eats plants
Orthoptera

Mole cricket
Gryllotalpa sp.

The large, shovel-like fore legs are the most distinctive feature of mole crickets. As you will know if you have picked them up, these fore legs are exceptionally strong and can force apart your thumb and forefinger as you hold them tight. Mole crickets use these legs for digging their burrows, and quickly disappear into the soil if disturbed.

40 mm

Eats plants

Orthoptera:
Gryllotalpidae

Mole crickets have permanent burrows but from these extend temporary feeding galleries. Normally they feed on plant roots, but some species are suspected of eating soil insects as well. Mole cricket burrows are homes where they lay their eggs, and then care for the brood. You usually find mole crickets when digging in the garden, or when burrows just under the surface produce bulges in otherwise smooth turf. They are often found above ground after rain, possibly because their burrows are flooded.

After summer rains mole crickets can be heard chirping away (until you get too close). They produce sound by rapidly rasping their wings together. However, the cunning mole crickets have worked out that the sound they produce can be amplified significantly by constructing a special shaped entrance to their burrows. It is so loud that it actually begins to hurt your ears if you can sneak right up to a singing mole cricket. In fact, you are hearing amplified music from your noisy insect neighbours.

Another type of cricket, the sandgroper (in the family Cylindrachetidae), looks like a mole cricket but is actually a distant relative. (Sandgropers are truly insects, not just Western Australians.)

Grasshopper
Acrididae

Grasshoppers, like crickets, have large hind legs used for jumping. However, they have much shorter antennae than crickets. One well-known type of grasshopper is the plague locust. *Locusta migratoria*, the migratory locust, is a major pest overseas, but the Australian plague locust, *Chortoicetes terminifera*, is our serious problem.

It is important to remember that not all grasshoppers are locusts. There are over 700 species in the grasshopper family Acrididae in Australia, of which only a handful of species form pest locust swarms. Wingless grasshoppers (*Phaulacridium vittatum*) can be a more regular problem in gardens in summer, as the grasshoppers look for green plants to eat when pastures dry out.

Young grasshoppers look like smaller versions of the adults, but only the adults can have fully developed flight wings. In some species, such as *Phaulacridium vittatum*, most adult individuals have very short wings. But when conditions change (and food becomes in short supply) more fully winged individuals develop. Some desert and alpine species of grasshoppers always have tiny wings as adults, as they do not want to fly. Flying would only take them away from the small area that is suitable for them, so they have done away with the redundant flight wings.

20 mm
Eats plants
Orthoptera

Spurlegged stick insect, spurlegged phasmatid
Didymuria violescens

While you are likely to see stick insects, you will see them infrequently. They generally live quite solitary lives and even distance themselves from other members of the same species. Most Australian species are wingless, although the adults of some species living in trees are winged and can fly. Most stick insects are slow moving. Even if you catch one it will not always take flight. They rely on camouflage for protection and, if disturbed, will often sway rather than run away.

Immature stick insects look very much like the adults, except for the species that have wings. The hind pair of wings are the flight wings, and these are folded under the fore wings, which act as protective covers. Both adults and immatures feed on leaves. The youngest individuals feed on the softer, younger leaves at the tips of branches, while older individuals can feed on both young and old leaves. The eggs are ornate caskets that look like seeds, as you can see in the photograph.

There are about 150 species of stick insects in Australia, of which only three species are considered pests. The pest species can occur in larger numbers than most and feed on eucalypt trees. They can cause defoliation of commercially valuable trees and so become pests.

One of the largest insects in Australia is a stick insect (*Acrophylla titan*), with a spiny body that can be up to 25 cm long. An unusual feature of some stick insects is their readiness to shed limbs if attacked, and then the ability to regrow whole limbs if they are lost.

3 mm (egg),
95 mm (adult)
Eats plants
Phasmatodea

Psocid
Psocoptera

Psocids are tiny insects that look quite like lice, at least until the adults of some species develop wings. They are often found in damp places, feeding on mouldy objects including damp papers and books, which is why some species are called booklice.

Psocids are usually drab coloured, and those with wings hold them tent-like over their bodies. They also commonly occur in garden mulch, on dead leaves or branches, on damp wallpaper, on fence palings and in stored, floury food. You can find a few species in dry environments but most prefer damp surroundings. Many winged species are found on trees while the wingless species occur in more sheltered or closed environments.

Some species make a tapping or ticking sound by hitting their abdomens against whatever they are standing on. The noise is thought to be the mating call of some female psocids.

Outdoors they are a key link in the food chain. They are small enough to feed on microscopic fungi and moulds, and big enough to be food for other insects, spiders, birds and lizards. Apart from the few pest species infesting food, psocids in houses are simply a nuisance. They can be found trapped in sinks and bathtubs, unable to climb out.

Psocids are also pests in insect collections, especially if kept in a damp room. So watch out for psocids if you start a collection!

3 mm
Eats detritus
Psocoptera

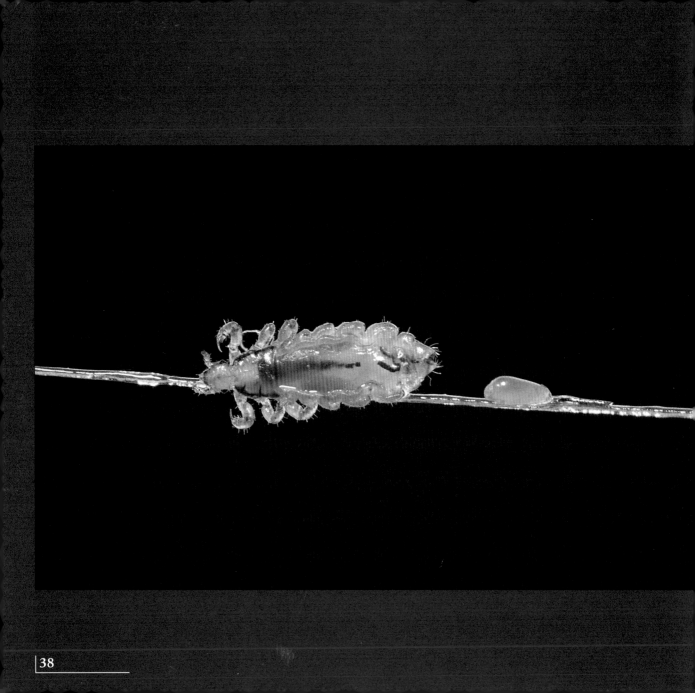

Head louse
Pediculus capitis

Lice! The very word makes people begin to itch. Lice are a very specialised group of insects that have adapted to live as parasites on birds and mammals. It is estimated that over 2500 species of lice occur in Australia, but only three species (head lice, *Pediculus capitis*, body lice, *Pediculus humanus*, and crab lice, *Pthirus pubis*) have adapted to live and feed on humans. Until recently head and body lice were considered to be variants of the one species, *Pediculus humanus*.

The bodies of lice are flattened (top to bottom) and their legs are adapted to grip the hairs or feathers of their hosts. Lice do not have wings, and rely on close contact between hosts (for example, in a schoolyard) to spread. Head lice have a life cycle of about twenty days, but adult females can live for over thirty days and lay up to 100 eggs or 'nits' (you can see a nit in the photograph). Each egg is cemented on to a hair with a glue so strong that nothing safe to humans can dissolve it.

Lice can carry epidemic typhus (not known in Australia for many years). They are suited by conditions where people cannot wash themselves or their clothes regularly (and remove lice), and where conditions are crowded (so the lice can transfer easily between hosts). Wars create ideal conditions for lice and the first major use of the insecticide DDT was to control lice on humans in World War II. Lice can develop resistance to insecticides routinely used against them, and the use of a warm, fine comb remains a good control measure.

Lice are useful for zoologists studying animal distribution and evolution, as parasitic lice often develop into new species faster than their hosts.

2.5 mm
Parasite
Phthiraptera

TRUE BUGS
Hemiptera

MOST PEOPLE CALL INSECTS and other invertebrates 'bugs'. It comes as a surprise and, we suspect, a disappointment when an entomologist tells them that their specimen is a bug. Yet there is an order of insects, the Hemiptera, that are true bugs.

The group is a very large one, with many commonly found species including aphids, scale insects, cicadas and bed bugs. All species of true bugs feed by sucking, not chewing, and this is one of the characteristic features of the order. There are three major parts to the order, based largely on different wing structure and body form. Examples of the three groups are scale insects and aphids (have few veins in wings and are generally soft bodied); leafhoppers and cicadas (have harder bodies and wings with many veins); and crusader bugs and assassin bugs (are hard bodied and have fore wings with a toughened base).

There are over 5500 Australian species of Hemiptera known and more are certain to be found. There are many important pest species, but equally there are many important beneficial species.

Mating pairs of harlequin bugs walking around attached to each other are a common sight during summer in southern Australia. Some other bugs, such as bed bugs, have a much more bizarre method of mating. In these species the male punctures the body wall of the female, and the sperm he deposits migrate to the egg through her body fluid (haemocoele). This apparently brutal method of mating is successful but restricted to relatively few species!

Lerp insect
Glycaspis sp.

The term 'lerp' really refers to the case covering an insect in the family Psyllidae, although it is sometimes wrongly used to describe the insect as well. Most of those we see feed on gums (*Eucalyptus*) and wattles (*Acacia*), with many individuals fitting on a single leaf.

The insects suck sap and have to filter a lot of fluid to get a meal, which means that they excrete a sugary honeydew as waste. Many lerp insects use the honeydew as the base for their hard protective covers. These covers can be simple or very ornate (such as the lace lerps, *Cardiaspina* species) depending on the species as each has its own characteristic design.

Large numbers of lerp insects sometimes develop, causing damage on either individual trees or on a larger scale throughout a district. Plants that have already been stressed in some way, for example by drought or by growing in an unsuitable place, are particularly vulnerable to lerp damage. However, lerp insects have many natural enemies that keep them largely under control, including both insect predators (such as lacewings) and vertebrate predators (such as birds).

Lerp insects have been on the menu of humans for many centuries. Aborigines ate them for their sweet sugary value, but in biblical times it is likely that the manna that saved the Israelites was dried honeydew from another hemipteran.

10 mm (lerp)
Eats plants
Hemiptera:
Psyllidae

Scale insect, rice bubble scale
Eriococcus coriaceus

At first glance these insects look like galls on the young growth of eucalypts. While some scale insects do induce galls, this species does not. The globular lumps are females, often found in lines or clusters on the twigs.

Females lay eggs underneath themselves and use their bodies to shield their young until they can emerge and find feeding sites. Males are smaller and often found in separate clusters. Only adult males are winged but, perhaps in exchange for that privilege, cannot feed. As a result, they are very short-lived and barely have time to mate before dying.

The adult males are weak flyers and cannot move far without help from the wind. The youngest nymphs ('crawlers') are so light that they are carried along by wind currents. New colonies are started when crawlers move along to new shoots or drift to trees further away. Once the nymphs have moulted and begun to grow, they lose the ability to move. They more or less glue themselves to a branch, where they are vulnerable to predators and parasites.

Scales have an unusual defence: ants. Like aphids and lerp insects, scales produce honeydew as a result of feeding on sap. Ants collect the sugary waste and, in turn, keep invertebrate predators and parasites away. While many scale insects are farm or garden pests, scales of the genus *Dactylopius,* commonly called cochineal insects, produce the red food colouring cochineal.

5 mm
Eats sap
Hemiptera: Coccoidea

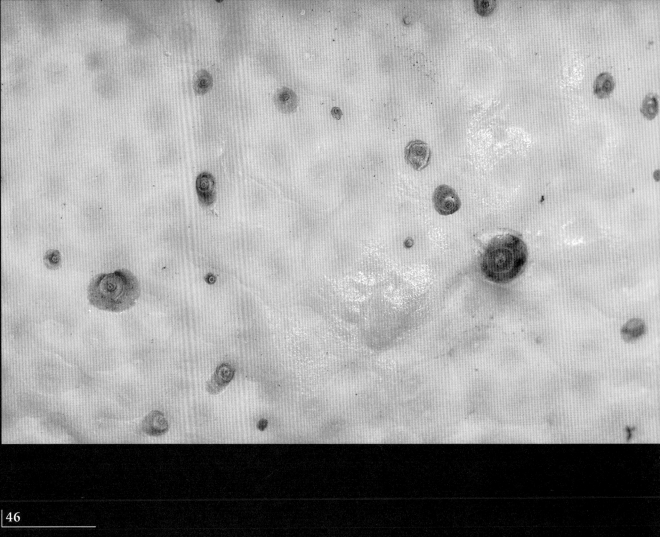

46

Red scale, hard scale
Aonidiella aurantii

When you look at this species, it is easy to see why they are called *scale* insects. Where heavy infestations occur, the insects really do form a scaly covering over the plants they feed on. In fact, they don't look like insects but just waxy little specks on fruit or plants. Insects like this are commonly called hard scales.

Red scales are very common on citrus trees, and are often seen on lemon, orange and mandarin trees. They begin life as 'crawlers'. The female does not lay eggs but instead produces active nymphs. The nymph can move around and find a good place to feed. After it finds such a place it produces a hard waxy covering and remains there. As with many scale insects, the crawlers can be blown quite a long way by the wind and so populate different areas. However, if a plant is a good host, many crawlers will simply move a short distance to younger shoots or fruit on the same tree or bush.

In common with other scale insects, females and nymphs look alike, while adult males have wings and can fly weakly. Black sooty-mould is often associated with scale insects as it also grows on the sugary deposits left on leaves.

Control of this pest species in orchards is one of the success stories of modern pest management. Growers either release or simply encourage its natural enemies, including minute wasps, ladybirds and lacewings. Commercial insectaries produce and sell these beneficial insects as an alternative to conventional chemical insecticides.

2–3 mm
Eats sap
Hemiptera:
Diaspididae

Aphids
Aphididae

There are very many species of aphids that commonly occur in urban Australia. The two aphids shown here, the green peach aphid, *Myzus persicae* (opposite), and the rose aphid, *Macrosiphum rosae* (next page), are common but they are not native species. We usually see them as pests on several introduced plants such as roses, peaches, nectarines and potatoes. In large numbers they can damage or weaken plants by draining fluid from them. Green peach aphids also carry and transmit several serious plant viruses.

These two species do not like hot dry Australian summers, but will prosper in warm, humid conditions. They have a very short life cycle (as little as seven–ten days), and give birth to live young (as the aphid in the opposite photograph is doing). Consequently, they seem to appear from nowhere in plague proportions. In fact, a small population can change to a large one between weekends!

Aphids have very complex life cycles, which are affected by the weather, population density, the time of year and the state of their host plant. Adults may be either winged or wingless. After winter winged aphids generally migrate to new plants where they begin new colonies. As the new host plant will usually exist for a few months there is no need for the insects to move far—and no need for them to have wings. Wingless adults are the norm until the population is too large or the plant is dying. Then, wings are required again and the next generation of (winged) aphids flies away to different hosts.

Aphids are true bugs, and so have sucking mouth-parts. They suck sap and, like scale insects, produce honeydew as a waste product. Ants love to collect honeydew and will

2 mm
Eats sap
Hemiptera:
Aphididae

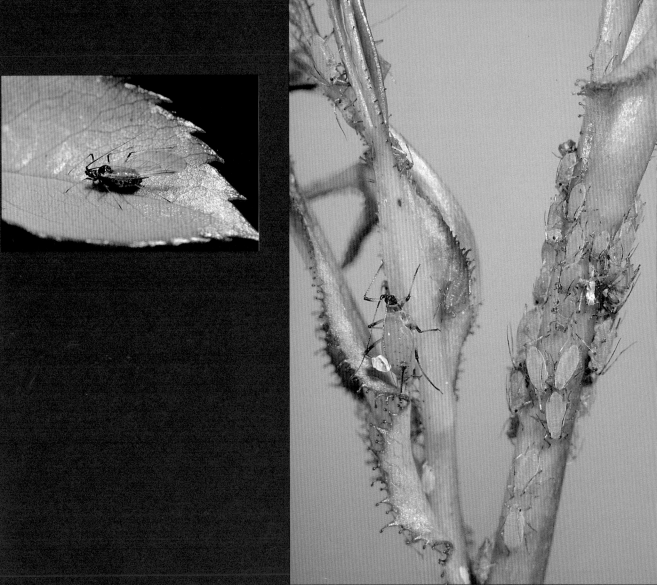

protect aphid colonies in order to keep the production going. Despite the protection of ants, aphid colonies are almost always destroyed by either predators (ladybirds and lacewings) or parasitic wasps. The parasites live inside the aphids, puffing them into round 'mummies' before cutting a neat hole from which the adult wasp emerges.

Spittle bugs
Cercopoidea

Spittle bugs are commonly found in two forms, neither of which look at all like insects. There are species that envelop themselves in froth, 'spittle' (right photograph), and those that encase themselves in conical tubes (left photograph). These cases protect the nymphs, and they hide the insects very effectively. When the adults emerge, they look like very small cicadas.

In both groups, the immature spittle bugs live in what we would consider an awful situation: a liquid environment of their own excreta. But the spittle bugs obviously like it and it must make them very difficult for their natural enemies to find.

The frothy spittle bugs are more obvious, and you will often see large blobs of white froth on small bushes or weeds. The conical tube forms are less obvious, but you can see them easily on the younger parts of plants, especially eucalypts or acacias.

4 mm (insect in froth), 12 mm (tube)

Eats sap

Hemiptera: Cercopoidea

Greengrocer cicada
Cyclochila australasiae

This common cicada is well known to most people in south-eastern Australia because of its loud song on hot summer days and evenings. Cicadas of this species naturally feed on several types of eucalypts, but will also feed on many other plants including introduced species. (Another photograph of the greengrocer cicada appears on p. 40.)

Adult female cicadas cut slits in the young shoots of plants, into which they lay their eggs. After hatching the tiny cicada jumps from the plant to the ground. It has no wings and relies on its small size to drift down without injuring itself. It burrows into the soil and will spend most of its life below the ground sucking on the roots of its host plant, moving very little. When conditions are right, the nymph constructs a tunnel to the surface. It climbs up the stem of the plant, sheds its skin and a fully winged adult emerges. We know very little about the duration of the life cycles of Australian cicadas. Usually we see only the moulted skins and the noisy adults. Some, such as the greengrocer, are thought to live for six or seven years below the ground. Many other species probably have life cycles of one or two years. One American species emerges only every seventeen years.

Cicadas come in many different colours and patterns, including those with bright red and orange markings as well as black, brown and green. There are also tiny species less than 10 mm long that live on grass, in addition to the large, obvious species that feed on trees.

55 mm
Eats sap
Hemiptera:
Cicadoidea

Assassin bug
Gminatus australis

As its name suggests, the assassin bug is a killer. There are many species called assassin bugs—the one in the photograph is *Gminatus australis*—but all have the same manner of feeding. The sucking mouth-parts of a true bug are modified in assassins to become a weapon (proboscis) with which to kill other insects. The assassin bug is typically slow moving and ambushes its prey. It grasps its victim with its legs, punctures it with its powerful proboscis and then sucks out the body fluids.

Assassin bugs are important in controlling pest, and potential pest, species. They will feed on other insects such as plant-feeding bugs and prevent damaging numbers of pests from building up. This is particularly so where the pest may lay a batch of eggs, or a group of nymphs develop together in a cluster. The predatory assassin bugs will feed on one pest after another and so clean out an infested area.

A word of warning: the role of assassin bugs as friends in our gardens is quite clear but you need to be cautious if handling them. They can bite people just as well as they can bite other insects and this can be painful. Overseas, several species feed on human blood, but these insects have not been found in Australia.

11 mm
Predator
Hemiptera:
Reduviidae

Crusader bug
Mictis profana

The conspicuous yellow-orange cross on the back of this bug and the bug's fairly large size (about 2.5 cm) make it easy to spot. Like all bugs it has sucking mouth-parts, and this species uses them to extract sap from plants such as acacias, cassias and citrus.

Crusader bugs are found in many parts of Australia and also on some Pacific Islands. They attack plants, but they are not usually found in large numbers. In the home garden it could be controlled by hand in most cases, or you could simply tolerate them. It is not known to spread any particular diseases.

25 mm
Eats sap
Hemiptera

Harlequin bug
Dindymus versicolour

The name 'harlequin bug' means different things in different parts of Australia. The harlequin bugs in the photograph are found in South Australia, Victoria, Tasmania and New South Wales. This species is a true bug, and sucks sap from a range of plants, including common garden plants (including fruit and vegetables) and weeds. In addition to feeding on plants, harlequin bugs will also suck juices from the carcasses of animals (for example, birds, mammals, insects and snails). The name comes from the colourful adult body, with red, black, yellow and green in large patches.

12 mm
Eats sap
Hemiptera

The adults are often found as mating pairs, giving rise to other colourful names used for this species. A mating pair attach themselves at the tips of their abdomens with a ball-and-socket joint that allows them to move relatively freely. The larger of the pair usually drags the smaller one along behind.

Nymphs have fewer colours, but are still very bright with red wing-buds. They are usually found in large numbers with the adults.

Both nymphs and adults like to retreat into sheltered sites such as leaf mulch, vegetable compost, fence palings and hedges. Harlequin bugs are most abundant in the warmer months, although you can find active adults on sunny days in winter.

Green leafhopper
Siphanta sp.

There are several families of bugs that are commonly called leafhoppers or plant-hoppers, and this is one of the larger, more obvious species, belonging to the genus *Siphanta*. Many leafhoppers are small, torpedo-shaped insects, often with striped abdomens. This one is quite different as it has broad, triangular wings that it holds vertically.

The large wings may seem conspicuous, but on a plant they look much like green thorns. Both adults and nymphs can jump away if disturbed, but in most cases they will first try to avoid whatever is worrying them. They do this by moving to the far side of the stem. It is quite amusing to watch when a line of green thorns moves at once!

The nymphs produce and cover themselves with white wax, making them look as if they have a white coat and long white tails. The nymphs do not like to move and, if they do, some of the white wax remains behind on their food plant after they leave. The wax helps to protect the nymphs from predators, perhaps making them look larger than they are.

10 mm
Eats sap
Hemiptera: Flatidae

Thrips
Thysanoptera

Thrips are slender, tiny little insects with adults having narrow wings fringed with hairs. Scientists have found over 400 Australian species and more species are certain to be discovered. However, it is the plague thrips (*Thrips imaginis*) that most people notice. It is a yellowish insect that feeds in flowers and becomes abundant in summer. The numbers of thrips can become so high that flowers are killed and, in fruiting plants such as apples, the crop lost. The plague thrips (an individual is a 'thrips' not a 'thrip') is also fond of landing on washing hanging on clothes-lines.

Thrips are closely related to true bugs (Hemiptera) and have sucking-type mouth-parts. Because thrips are so small, they can pierce individual plant cells and suck out the contents. After feeding they characteristically leave a dark dropping, like those on the left of the photograph.

Individual thrips are usually short lived (a few weeks or so) and have an unusual life cycle. After hatching from eggs, they develop into larvae but then pass through one or two resting stages similar to the pupal stage in the more advanced orders of insects. Then the adults, which may or may not have wings, emerge.

Although only a few millimetres long, some species of thrips can travel great distances. One species is thought to fly regularly to New Zealand from eastern Australia. Obviously they are more carried by the wind than by their own wings but, as a species, they regularly succeed in making the journey.

1.5 mm
Eats plants
Thysanoptera

Green lacewing
Chrysopidae

The beautiful, golden-eyed green lacewing is an insect that you will often see at your windows at night, as it comes to lights.

Both adult lacewings and the larvae are voracious predators of aphids and other soft-bodied insects (such as caterpillars). The lacewing in the photograph appears to be sizing up the aphid on the other side of the stem. Commercial insectaries in Australia and overseas rear this species, *Mallada* * *signata*, and other lacewings for use against pests instead of chemical insecticides.

The larvae look like insect versions of crocodiles. They have large jaws projecting in front of them to seize any prey. To help them sneak up on their unsuspecting prey, and also to avoid becoming food for larger species, they camouflage themselves—in many cases with the remains of their previous meal (as the larva in the photograph has done).

The adults are easily recognized by their 'lace' wings, formed by branching veins. Many green lacewings lay their eggs singly on the end of long stalks (as in the photograph) mainly to prevent the first-hatched, hungry larva from eating all the other eggs! These stalked eggs often remain after the larvae have hatched, and can be found attached to leaves or buildings.

The larvae of another group of lacewings, called ant-lions, dig pits to catch small insects such as ants.

1 mm (egg),
5 mm (larva),
15 mm (adult)
Predator
Neuroptera

---

* Many of the common green lacewings were until recently called *Chrysopa* species. You may find them listed in textbooks under that name.

BEETLES
Coleoptera

I F SUCCESS IN THE ANIMAL WORLD is measured by how many species there are, then beetles are the most successful of all. There are over 300 000 species, and about 30 per cent of all animal species in the world are beetles. A famous biologist once remarked that God must have had 'a special preference for beetles' as there are so many.

Entomologists estimate that there are about 30 000 species of beetles in Australia. Beetles have been able to inhabit almost every type of habitat from freshwater to dry land and some are even parasitic. One reason for their success is the way that they have modified their wings to provide protection as well as flight.

All winged insects except flies have two pairs of wings. In beetles, the first pair of wings has become a tough covering (called the 'elytra') for the abdomen and the hind pair of wings. This cover acts like armour and helps beetles to inhabit environments that would otherwise be too harsh. They are less vulnerable to predators than soft-bodied insects.

A drawback with the wing cover is that it is more difficult for beetles to fly. The covers must be opened, the hind wings must be unfolded and then of course there is only one pair of flight wings instead of two. As a result, beetles fly more like helicopters than jets, but the advantages obviously outweigh the disadvantages.

Ground beetle
Carabidae

This fast-moving, black beetle, *Notonomus gravis*, lives, like most carabids, on the soil surface (which is why carabids gain their name of ground beetles). This carabid does not fly as its flight wings (hind wings) are tiny and do not work. In fact, its fore wings are fused together to form the elytra. It catches its food by chasing insects on the ground or scavenging dead or dying individuals. Both adult beetles and immature grubs (larvae) are predatory, but the larvae normally feed below the ground. Adults may be found all year round but are most common in summer and autumn. The adult beetles suddenly appear in large numbers at about Christmas or New Year, after emerging from their pupal stage. The female of this species has a behaviour that is rare in ground beetles; she cares for her egg brood in a burrow until the larvae hatch and disperse. This takes many weeks, and during this time the mother probably goes without food.

This species typically inhabits grassland on Victoria's western plains, coming as far east as the eastern Melbourne suburbs. However, the genus *Notonomus* is large, containing over 100 species. These are found along the ranges of eastern Australia from Queensland to Tasmania. Adults can be found sheltering under logs or rocks; larvae are difficult to find as they usually burrow deeper in the soil. Like most carabids, adults have a pair of defensive glands on the tips of their abdomens that they use to squirt out nasty-smelling chemicals.

Carabid beetles attack many other insects, including species we regard as pests, and this has led to them being considered beneficial in agricultural ecosystems. *Notonomus*

18 mm

Predator, scavenger

Coleoptera: Carabidae

gravis too may act in this way in the pasture lands of Victoria from Melbourne to the Grampians, as it likes to eat many common caterpillar pests.

Other carabid beetles, such as *Chlaenius* species, are much more colourful (often metallic) and can fly. Species of *Chlaenius* are widespread in Australia. *Chlaenius australis* is found in every state and the Northern Territory, and *Chlaenius darlingensis* (shown in the photograph) is commonly encountered in all states except Tasmania. Other widespread and brightly coloured carabid beetles are in the genus *Calosoma*. Species in this genus are found in many countries and, although we have only three species in Australia, they are noticeable (being bright green) and are voracious predators of caterpillars.

Rove beetle
Staphylinidae

At first glance, rove beetles do not really look like beetles at all. The main identifying feature of beetles, the elytra or hard wing-covers, appears to be missing. Adult rove beetles look more like the nymphs of other insects, such as silverfish. In fact the elytra are there, just very shortened.

The short elytra leave most of the beetle's unusually flexible abdomen exposed. As a result they seem to slither through grass, or around whatever is in their path, trailing their abdomens behind them. Under their very short elytra, the rove beetles keep their full-sized flight wings. They have a very complicated folding procedure to get them neatly stowed away, but they do this quickly and easily. They can unfold these wings and be airborne in a moment.

When disturbed a rove beetle can arch the tip of its abdomen back over its head. Some of the larger species give off a strong smell to deter other animals that may want to eat them. Some species of *Paederus*, distinguished by their bright orange markings, have another defence. They release a liquid that causes blistering on the unwary humans who pick them up. The wounds this causes has lead to them being called whiplash beetles.

Not all rove beetles are predatory, but many of the larger species prey on other insects. The devil's coach-horse, *Creophilus erythrocephalus*, feeds on maggots and has a black body and bright red head. One genus (*Cafius*) preys on insects and other small animals found in seaweed washed up on beaches.

8 mm
Predator
Coleoptera

Longicorn beetle, longhorned beetle
Cerambycidae

All adult longicorn beetles have very long antennae. While some have antennae much longer than their bodies, all have antennae at least two-thirds as long as their bodies. They are usually long, thin insects, like the species shown here. Many adult longicorns feed on pollen, but others are known to eat leaves or bark.

Almost all longicorn larvae feed on wood, tunnelling their way in trees or bushes. Females lay their eggs under the bark of the plants that they feed on. The newly hatched larvae can then tunnel straight into the tree or bush and avoid being exposed to predators or parasites. Small larvae are usually found just under the bark, spending the early part of their lives near the surface. Only later do they tunnel deeply into the wood. The larvae are usually legless, cream coloured, and long and thin. They have a broad head and thorax (the part of the body just behind the head) and large jaws. If you split firewood then you will often find them or their tunnels in the logs. The larvae turn into pupae in the tunnels, near to the surface so that when the adults emerge they can easily reach the outside.

Adults of many species look the same, with mottled brown and yellow markings. They are good flyers and often come to lights at night. With their hard bodies they make more noise at the window than moths!

25 mm
Eats plants
Coleoptera

Darkling beetle
Tenebrionidae

This darkling beetle belongs to the genus *Adelium*. There are many species of *Adelium* in Australia, and they all look very much alike. The adult beetles are black, and have a slightly dimpled or beaten look. The larvae are creamy-yellow coloured, and look a little like short worms. The larvae of some species of this group of beetles are bred for use as animal food and known as 'mealworms'. Others are agricultural pests and are commonly called false wireworms.

The larvae are often found under logs, where they feed on rotting wood. The adults shelter under logs too, but are also found sheltering under bricks or stones. The adults emerge from under their shelters at night to feed above ground on plant material. When disturbed, an adult beetle often protrudes a pair of yellow glands from the tip of its abdomen. This is a means of defence, and, as well as surprising a potential enemy, the glands produce a chemical repellent.

Apart from the agricultural pest species, we know little about the biology of most tenebrionid beetles. Even accurately identifying species is difficult because of the similarity of features among species in the group. So far, entomologists have found more than 1500 Australian species of Tenebrionidae and we are certain to find more!

15 mm
Scavenger
Coleoptera

Click beetle
Elateridae

Click beetles must be among the favourite insects of children. These beetles will arch their backs and leap into the air with a clicking sound, like an insect version of a wind-up toy.

Click beetles have a characteristic shape, like an elongated oval. On each side of their backs, on the part of the body between the head and the wings (the pronotum), they have points projecting towards their tail. On their undersides they have the clicking mechanism. This consists of a stiff rod that acts like a spring. When the beetle is disturbed it arches its abdomen and pronotum, and the movement releases the spring. The force can launch the beetle into the air and, with the associated clicking sound, this would surely frighten a potential predator.

The larvae usually live in the ground. Some larvae are pest species that eat germinating seeds and they are commonly called true wireworms. (False wireworms are in another beetle family, the Tenebrionidae). Usually these pests are found in crops such as wheat and barley, but some others attack root crops like potatoes.

Most click beetles have life cycles of a year, but some can live for five years or more as larvae in the ground before pupating and turning into adult beetles.

The adult beetles are often seen in summer, flying to lights at night. Some are large enough to make quite a noise when they hit your window!

10 mm
Eats plants
(most species)
Coleoptera

Christmas beetle
Anoplognathus sp.

The name 'Christmas beetle' means different things to different people, but usually it refers to a beetle in the family Scarabaeidae (see also p. 97) with a shiny or metallic body that we see around Christmas time. Several species in the genus *Anoplognathus* are known as Christmas beetles and they are serious pests. They eat leaves of eucalypt trees and, when present in large numbers, can defoliate or seriously damage trees. In some parts of Australia, such as the New England region in New South Wales, they have seriously damaged large numbers of eucalypts. Elsewhere they can damage young trees by attacking the new growth.

Christmas beetles have particularly long, curved and strong claws (you can see these clearly in the photograph on p. 68). No doubt these help them to grasp the edge of a gum-leaf while they eat it.

The larvae live in the soil and feed on the roots of plants. They too can be pests, because they can kill lawn grasses by eating them from beneath the ground. On the positive side, these grubs are a good food source for magpies!

25 mm
Eats plants
Coleoptera

Passalid beetle
Passalidae

There are only about forty species of Australian passalid beetles and all have a very similar body shape. They are large, black, shiny, solidly built beetles that are found with rotting logs or wood. The head of the adult beetle is often adorned with a short horn and always with a pair of large, curved antennae. The wing covers (elytra) over the back of the beetle are distinctively furrowed.

Passalids have a basic social behaviour, including audible communication. If you pick up an adult passalid it will usually move its abdomen so that specialized structures (like a guitarist's plectrum) rasp against the wing covers, producing a chirping noise.

The larvae are distinctive, apparently having only four legs. The third pair of legs exists only as stumps, but these are used to make rasping sounds. The reduced legs are more important to passalids for communication than for movement.

Both adults and larvae feed on wood. Their ability to process fallen timber by chewing it (therefore creating another food source for smaller creatures and liberating nutrients held in the timber) makes them a vital link in the food chain. They are particularly important in northern tropical Australia.

30 mm
Eats wood
Coleoptera

Fiddler beetle
Cetoniinae

Beetles are characterized by the hardened fore wings that form a protective covering (the elytra) over their backs. While this gives them protection it makes it more difficult to fly. Typically the fore wings are held out in flight and the hind wings do the work. The Cetoniinae are one of several groups of beetles that have developed a more efficient way of flying.

Beetles in the Cetoniinae group (a 'subfamily' in the family Scarabaeidae) have a characteristic side-cut to the elytra. This allows them to use their hind wings while barely raising their fore wings. As a result, these beetles can fly faster and more accurately than most others.

In Australia, as in Europe, these beetles are most commonly seen feeding on flowers, especially roses. However, they seem to prefer native Australian flowers and are not notorious garden pests here. The larvae live and feed in rotting wood.

These beetles are most common in the mainland eastern states. They are often marked with beautiful patterns and look as if they have been expertly hand-painted. The fiddler beetle, *Eupoecila australasiae*, with lyre-shaped markings, is one of the most well known.

17 mm
Eats plants
Coleoptera

Weevils
Curculionidae

You can immediately recognize a weevil as a beetle with a long 'nose', and a pair of elbowed antennae placed on it. Weevils are among the toughest of beetles, many having incredibly hard, spiky bodies. There are also a number of notorious pest species that are very hard to control.

Weevil larvae are different from those of most other beetles as they have no legs. Many also have tiny heads, making them look very maggot-like.

At least 6000 species of weevils occur in Australia but a few pest species are the ones we encounter most often. The pests attack garden plants (vegetables, bushes, lawn or trees) or stored food (grains, beans, flour and similar products). Introduced species such as the garden weevil (*Phlyctinus callosus*, pictured opposite), Fuller's rose weevil (*Asynonychus cervinus*) and the black vine weevil (*Otiorhynchus sulcatus*) are common pests in gardens and plant nurseries in Melbourne. The larvae live in the soil and feed on the roots of plants, sometimes ring-barking them, while the adults feed on leaves and young shoots. None of these three weevils can fly.

Around Sydney the native Botany Bay diamond weevil (shown on the next page) is well known. This spectacular species caught the eye of Joseph Banks, and so was one of the first Australian insects to be catalogued by Europeans and given a scientific name (*Chrysolopus spectabilis*). Its larvae feed on acacias and can cause serious damage to the plants. According to one source it is also known as the 'splinter puller'. If the weevil's head is pushed on to a splinter in your hand, it will clasp the splinter and pull it out!

6 mm (Garden weevil),
15 mm (Botany Bay weevil)
Eat plants
Coleoptera

Large weevils in the genus *Leptopius* are often found on acacias, which has earned them their common name of wattle pigs, but they also have a liking for apple trees where they can cause damage.

Ladybird
Coccinellidae

Ladybird beetles are, along with butterflies, one of the few types of insect that almost everyone thinks of as generally good. They have a characteristic oval shape and most species have bright-coloured patches or spots. Some ladybirds are not brightly coloured but they still have the same shape.

Both adults and larvae of most species, including the one illustrated, *Coccinella repanda*, are fierce predators of soft-bodied prey such as aphids. This is the main reason that they are seen as good insects, because, like lacewings, they help to control pests. The female beetles lay batches of bright yellow eggs on leaves near a suitable food source, usually near a colony of aphids. When the young larvae hatch they are quickly able to find some prey to eat.

One species of Australian ladybird, *Rodolia cardinalis*, was the saviour of the Californian citrus industry early this century. An Australian scale insect, the cottony cushion scale (*Icerya purchasi*), had become an established pest there and was out of control until the release of its natural enemy, the ladybird. The ladybird's subsequent control of the scale was one of the first, and best, examples of classical biological control.

There are a very few species of ladybirds that feed on plants and they can be pests in some areas. For example, the twenty-six spotted ladybird (*Henosepilachna vigintioctopunctata*) eats potato plants. The larvae eat the soft parts but leave the veins of the leaves behind, making odd patterns. However, such species are definitely in the minority. Almost all ladybirds are beneficial.

5 mm
Predator
Coleoptera

Plague soldier beetle
Chauliognathus lugubris

A beetle you may see in large swarms over summer is the brightly coloured plague soldier beetle. These beetles are unusual as they belong to a family (Cantharidae) that has very soft fore wings. They do not have the hard, shell-like elytra, or wing covers, of most beetles.

They are attracted in large numbers to a great range of flowering plants. They feed on the pollen and nectar, but also eat other, smaller, insects that arrive on the flowers. Their bright colours are a warning to potential predators that they taste horrible and can be poisonous. The larvae live on the soil surface where they are also predatory.

The adults are sometimes discovered on trees with damaged fruit. They will not have caused the damage but will be getting a drink from the wounds.

14 mm
Predator, eats pollen
Coleoptera

Cockchafer
Scarabaeidae

Cockchafers are the larvae of certain scarab beetles, some of which are illustrated in this book. They may be either pest or beneficial, but are almost always found in the soil. They are C-shaped grubs (sometimes called curl-grubs) with legs and a distinct head.

The pest species are those that feed on the roots (or come up at night to feed on the leaves) of plants including lawn grasses. Beneficial species include the dung beetles, which bury animal droppings and so prevent them becoming a breeding ground for flies.

5–20 mm
Eats plants, dung
Coleoptera

Jewel beetle
Buprestidae

It is easy to see why these insects are called jewel beetles. Their striking, often metallic, colours make them very beautiful insects. In some places they are actually made into or incorporated into jewellery. Here in Australia they are an obvious target for insect collectors and, in some states, jewel beetles are protected by law.

The beautiful colours that make jewel beetles vulnerable to humans are actually protection against animal predators. The strong, contrasting colours warn potential predators that the insect contains nasty tasting or poisonous chemicals. A bird, for example, would only have to eat one to know that it was not good food. Then all other jewel beetles of that species, and all that look similar, are safe from that bird! The bright colours make a clear warning that is easy to remember.

Most adult jewel beetles feed on nectar from flowers. They are most easily found on clusters of native flowers during spring and summer. The species pictured is a *Stigmodera* from the Adelaide Hills. There are very many species in this genus, found in many parts of Australia. Spectacular species of jewel beetles are particularly common in the heathlands of Western Australia.

Although the adults are brightly coloured and obvious, the immature beetles are not. The larvae usually live below the ground, feeding on the roots of plants. Sometimes they are known as witchety grubs, along with other beetle larvae that feed on roots, although this name is more correctly used for a particular type of caterpillar in the family Cossidae.

10 mm
Eats plants
Coleoptera

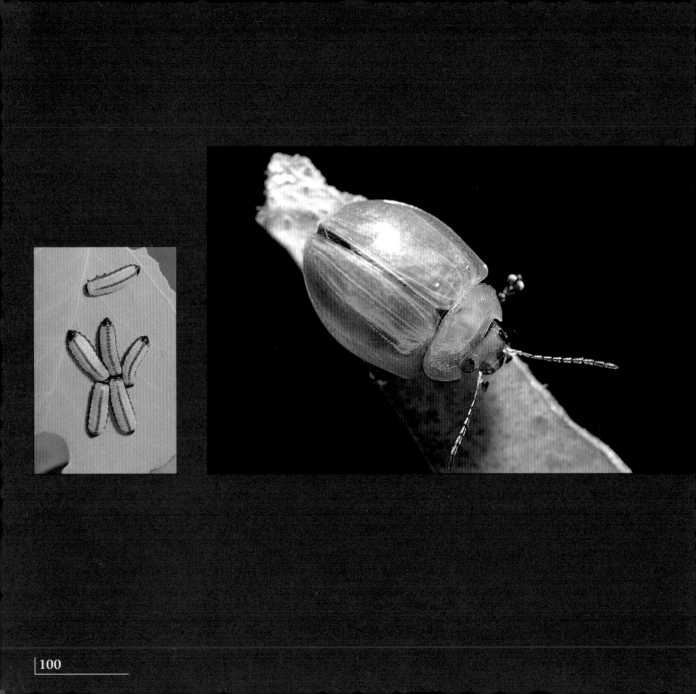

Tortoise beetle
Chrysomelidae

Chrysomelid beetles look like large ladybird beetles. Many are very round beetles, almost like a sphere cut in half. They are called tortoise beetles as they look as though they are well protected by a smooth shell. Two groups (genera), *Paropsis* and *Chrysophtharta*, are particularly common on eucalypt trees. A species of *Chrysophtharta* is shown in the photograph.

10 mm (larva)
8 mm (adults)
Eats plants
Coleoptera

The adults of *Paropsis* can be brightly coloured (sometimes vivid red), probably as a warning to potential predators that they do not taste nice. The larvae of beetles in this genus and related species make the poisonous compound hydrogen cyanide. Obviously it is risky eating cyanide, and so any predators would have to develop methods to cope with this form of defence.

Larvae of *Chrysophtharta* can be found in clusters on eucalypt leaves, particularly the younger leaves, in spring. The adult beetle lays clusters of yellow eggs which hatch into yellowish larvae with black heads. At this stage they can look like very young clusters of sawflies. As the larvae grow larger they tend to remain in a cluster. If you disturb them they will respond by arching their tails towards you, often with a drop of their defensive chemical at the end.

There are very many species of chrysomelid beetles in Australia, about 3000 at last count. Of these, a few are pests (in forestry, agriculture and horticulture) but most exist in balance with other species and do not usually form large, damaging populations.

Wood borer
Anobium, Lyctus

Wood-boring beetles, which can belong to one of several beetle families, are variously known as furniture beetles (*Anobium punctatum*), powderpost beetles (*Lyctus*), wood-worm, deathwatch beetles, auger beetles and shothole borers. They are all beetles whose larval stages attack dried timber.

The furniture beetle attacks seasoned wood, and so our furniture makes an ideal home for it. The first signs of furniture beetles are usually holes appearing on the surface of the wood as the adult beetles emerge (the holes in the photograph are up to 2 mm in diameter). If the wood surface is suitable, these adults can lay eggs on it and successive generations can develop within the same piece of timber. Eventually, the inner wood will be turned into a fine powder by the larvae, and the surface will look as if it has been blasted with fine shot.

Eats wood
Coleoptera

Museum beetle
Dermestidae

Known to many as carpet beetles, but to insect collectors as museum beetles, there are several species of dermestids that will attack animal-derived products: woollen carpets, furs, leather hides, feathers, or dried insects. The beetles are small and round, but the larvae are cylindrical and very hairy.

The presence of large numbers of these beetles in a house may mean an infested carpet, or may also indicate a dead bird or mammal in the ceiling!

3 mm
Scavenger
Coleoptera

Cat flea
Ctenocephalides felis

The fleas we know only too well are the cat flea (*Ctenocephalides felis*), the dog flea (*Ctenocephalides canis*) and the human flea (*Pulex irritans*). There are several thousand species of fleas in the world and many are associated with one or more warm-blooded hosts. As well as causing nuisance bites, fleas can also spread diseases. The most famous example is perhaps bubonic plague, spread by rat fleas, while a common example is murine typhus.

Adult fleas have thin, flattened bodies covered with backward-pointing spines to help them move through hair and fur and also to help them stay on. Like most fleas, cat, dog and human fleas feed on blood as adults, but the young insects are free living. The maggot-like larvae live in our carpets and dust, and in our pets' beds, not on animals.

Fleas can remain in a pupal cocoon for months until stimulated to emerge by factors including vibrations. This accounts for the almost yearly event of people returning home after summer holidays and, almost immediately, an apparent plague of fleas arriving. What actually happens in most cases is that many fleas remain in the pupal stage while the house is quiet and the disturbance caused by the returning humans triggers the simultaneous emergence of the adult fleas.

The life cycles of some fleas are extremely closely tied to those of their hosts. Reproduction in rabbit fleas is controlled by hormones from the rabbit host. That way, when young rabbits are born, a new generation of fleas is there to greet them!

1 mm
Parasite
Siphonaptera

FLIES
Diptera

I N AUSTRALIA, flies are things everyone knows about! And yet really we are only bothered by a handful of flies out of the more than 10 000 Australian species. Adult flies can be easily distinguished from all other insects as they have only one pair of wings. What would have been the hind wings are reduced to club-like stubs called 'halteres'. Wasps often look similar to flies but they have two pairs of wings and usually have the fore and hind wings coupled together so that they act as one.

Flies eat a vast variety of food; different species feed on plants, animals, fungi and rotting material, and they can even be parasitic. Adults and larvae of the same species usually eat quite different things. Common pests include the house fly, blowfly, fruit fly and March fly. This huge group of insects, which also includes mosquitoes, is of immense importance in our ecosystems as, in addition to the pests, there are thousands of important beneficial species, including those helping with decomposition, that fulfil a vital role in various food chains. They provide, if nothing else, links in the food chain that are critical for an ecosystem to function normally.

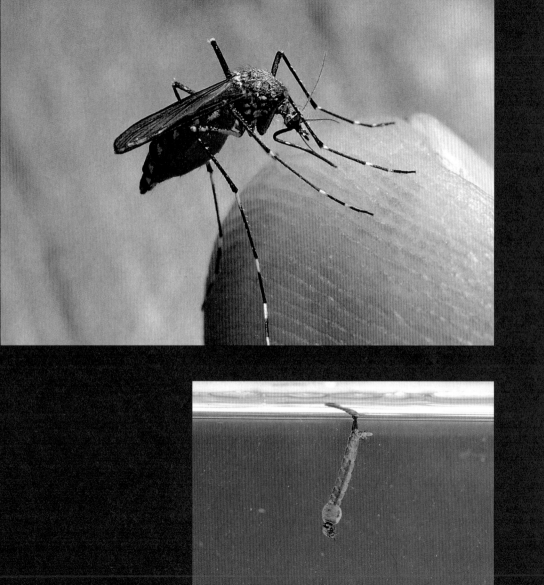

Mosquito
Culicidae

Mosquito larvae are aquatic, and are known commonly as 'wrigglers'. Although they live in water they breathe air by moving to the surface and breathing through a type of snorkel (called a siphon) fitted at the ends of their abdomens. They have special brushes on their mouths that filter food particles from the water.

The larvae develop into pupae (sometimes called 'tumblers' because of their curled shape), and then the adult mosquitoes that we all know emerge into the air. Only female mosquitoes suck blood; males eat sugary liquids such as nectar. Male mosquitoes can usually be distinguished from females by their bushy antennae. The familiar whine of a mosquito is produced when the stubby hind wings (halteres) vibrate rapidly.

There are several hundred different species of mosquitoes in Australia. They are important to us as a few species carry and transmit diseases. For example, species of *Anopheles* transmit malaria, and the common banded mosquito (*Culex annulirostris*) carries Murray Valley encephalitis and Ross River fever. Saltmarsh mosquitoes (*Aedes vigilax*) also carry Ross River fever, while *Aedes aegypti* carry dengue fever. However, most species that bite us are just a plain nuisance.

The adult females lay their eggs in an appropriate habitat for their species, but always on wet surfaces or on water. They often produce floating rafts consisting of many eggs cemented together. Even small water bodies, such as in containers holding plant pots, are suitable habitats for some mosquitoes.

7 mm (larvae),
8 mm (adult),
Eats detritus (larva)
Eats blood (adult)
Diptera

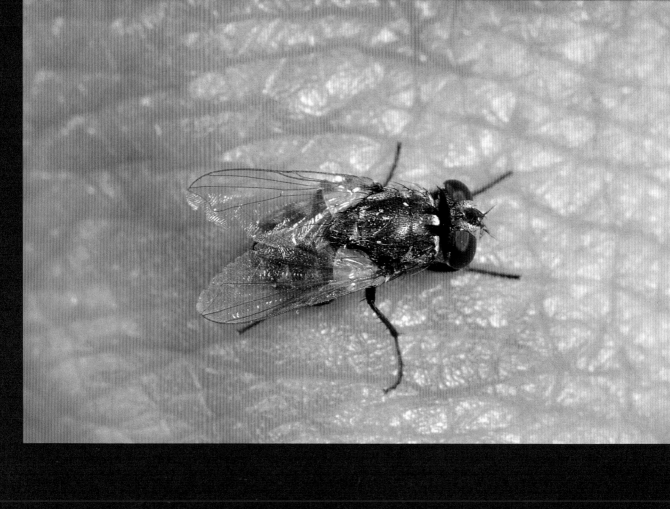

Bush fly

Musca vetustissima

The earliest-known written records of insect life in Australia describe the bush fly. Well before Captain Cook, Dampier, Torres and others documented the unpleasant company of bush flies in Australia. One explorer even named a place the 'Bay of Flies'.

Bush flies develop as maggots in the dung of many animals, including native marsupials and introduced sheep and cattle. They are about the same size as house flies, but they do not usually invade houses in large numbers. Adult bush flies have two stripes on their backs.

6 mm
Eats dung
Diptera

House fly
Musca domestica

Unlike the bush fly, house flies breed in grass clippings and similar compost, and invade houses. They are found in most parts of the world and are considered to be disease carriers. They can complete their life cycle from egg to adult in just over a week in good conditions. The immature stages are the well-known maggots and then a reddish-brown pellet-like pupa.

Slightly larger than the bush fly and with four stripes on its back, the house fly is well known to most Australians. It is abundant (in case you hadn't noticed) from spring to autumn.

7 mm
Eats detritus
Diptera

Blowfly
Calliphoridae

Lucilia cuprina, the most notorious of blowflies that causes fly-strike in sheep, is actually an introduced pest. (Fly-strike is the name commonly given to blowfly maggots eating living flesh on sheep.) There are other species of noisy flies that we call blowflies, including many native species, that belong to the same family, Calliphoridae. The photograph shows *Calliphora auger*, a common species.

As you have probably noticed, blowflies are attracted to meat or meat being cooked. They can lay maggots rather than eggs, so 'fly-blown' meat may have been exposed to flies for only a few moments. Unlike bush flies, they do not breed in dung. Most species of Calliphoridae normally breed in the carcasses of dead animals or (for some species) in living animals such as earthworms or sheep.

In many species, the adult flies generally have a metallic colour, giving rise to the common name of 'bluebottles' in addition to blowflies.

12 mm
Eats meat
Diptera

Tachinid fly
Tachinidae

These large, metallic, green flies are not large blowflies. In fact they are members of a family (Tachinidae), species of which are parasitic, mostly on other insects. Like parasitic wasps, the maggots feed on hosts such as caterpillars or true bugs. The maggots live inside the host insect until they are ready to pupate. The host normally dies and so these flies are important in controlling populations of potential pest insects. Adult flies most commonly feed on liquids such as nectar. Shown here is a species of *Rutilia*.

15 mm
Parasite
Diptera

March fly
Tabanidae

As you have probably found out, March flies appear well before March. There are many different species in this family (Tabanidae) and they appear at different times. Many have huge metallic-green eyes, which, as well as being quite striking, are probably very efficient at locating new victims!

March flies have a very straight, stout proboscis that the females of most species use to gather blood from their hosts. They wound their host and then mop up the blood as it arrives on the surface, rather than sucking it directly (like a mosquito). In other countries, March flies (also known as horse flies) can transmit human diseases, but this is not known to happen in Australia. However, some people can become sensitized to March fly bites and have a severe reaction if bitten.

The maggots generally prefer damp locations such as swamps or muddy ground, and so biting adults are often worse in such places. (The worst Paul has experienced was in the swampy tussock grass plains in Tasmania, during summer, while Denis has encountered hundreds on top of Mount Bogong in Victoria.)

13 mm
Eats blood
Diptera

Robber fly
Asilidae

Robber flies are strong, fast-flying insects that feed on other insects. They ambush other insects, often other flies, in the air. To achieve this they have powerful legs, covered with sharp spines to grasp their prey (you can see these spines quite clearly in the photograph on p. 108). Once they have captured their prey, they inject poisons into the animal to subdue it and then they secrete enzymes into the captured victim to help them to digest their meal. These enzymes break down the body tissues into a semi-liquid substance, which the fly sucks out leaving only the exoskeleton (cuticle) of the prey remaining. This method of feeding is very similar to that of predatory spiders (such as wolf spiders and jumping spiders).

Robber flies usually inhabit open woodland and so they would find many parts of cities where there are trees near open grassy land (such as parks, creek and river banks, and outer suburbs) quite suitable. Gardens with trees and flowers, which attract a range of insects that the flies can feed on, are especially suitable environments for them.

Robber flies often lay their eggs on the seed stalks of grasses, and the maggots develop in the soil or in rotting material. The maggots, like the adults, are often predatory.

23 mm
Predator
Diptera

Hover fly
Syrphidae

Hover flies are distinctive as adult flies because of their ability to do just what their name suggests—hover. They just seem to hang in the air before they dart away at great speed. They are seen in warm weather when flowers are open, and they certainly help to pollinate many plants. Adults of many species have yellow patterns on the abdomen, making them quite attractive, colourful insects. Their black-and-yellow colours make some people think they are stinging insects like bees or wasps, but they are not. Quite possibly their colouring disguises them and also confuses other insects and so helps to protect them.

What makes hover flies even more attractive for most gardeners is that the larvae (maggots) of many species are predatory. They love to feed on aphids in particular and so are very good to have around to control pests. It seems difficult to imagine a predatory maggot climbing plants, but this is just what they do. Fortunately for them, aphid colonies do not move quickly so the hover fly larvae can crawl up on their specially ridged and spined abdomens and attack the soft-bodied prey.

Other species have more normal maggot-like larvae, including the rat-tailed maggots found in drains and rotting, liquid material. The long tails of rat-tailed maggots are actually breathing tubes, which act like snorkels to allow them to live under (polluted) water.

9 mm
Predator
Diptera

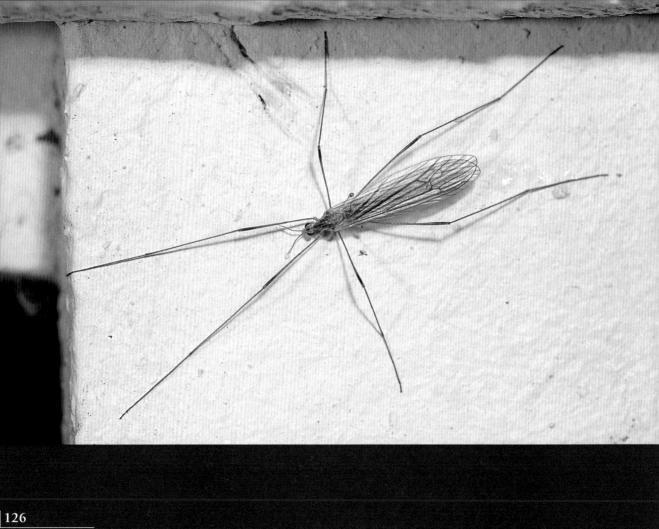

Crane fly
Tipulidae

Crane flies, also called daddy long-legs, are often believed to be immense mosquitoes. Thank goodness that they are not because, apart from their large size, there are more species of crane flies than any other family of flies. Over 700 native species have so far been identified in Australia.

Crane flies are most often found in the shady, damper areas of Australia. It is not only the adults who like wetter areas; the immature stages of many species are aquatic.

Many crane flies clearly demonstrate that flies (all Diptera) have only two wings. The hind wings are reduced to stubby 'halteres'. The halteres are easily seen in the photograph, looking like a pair of knobs on short stalks sticking out from the body behind the middle pair of legs.

10 mm
Eats detritus
Diptera

Moth fly
Psychodidae

These small, hairy flies have quite pointed wings that they hold over their bodies like a tent. Because of this they look like tiny moths. The moth fly in the photograph is a species of *Psychoda*. They breed in moist places such as drains (and sewage treatment plants) and are often found in damp places such as bathrooms. Their natural environment is shady creeks or swamps.

Moth flies have a very short life span of just a day or two. Females of other species in the same family (sometimes called sand flies) feed on blood!

2 mm
Decomposer
Diptera

Midge
Chironomidae

There are more than 200 different species of Australian midges known (so far). Most male midges have very hairy antennae. They can form large swarms, aggregating in a cloud-like group above some particular point they like, often in the evening. They may hover as a group over a particular bush or tree or in a clearing.

9 mm (larva), 5 mm (adult)
Eats detritus
Diptera

Like mosquitoes, midges have long, filament-like antennae. Although they may be a nuisance to people, they do not bite us. They simply occur in large numbers in some situations. Their close relatives the sand flies (Ceratopogonidae) and the black flies (Simuliidae) do bite humans and other mammals and those flies are serious problems in some areas.

Almost all midges have larvae that live in water and so they are more common near lakes, dams, rivers and other water bodies. Some are also capable of living in salty water and are found by the seaside. The larvae can live in the bottom mud of rivers and lakes, where there may be little oxygen and quite a degree of pollution. Some—like the one in the photograph—have the same red pigment (haemoglobin) as humans to help them carry oxygen and so they have become known as 'bloodworms'.

Recently, scientists have used midge larvae as indicators of pollution in rivers. Because larvae live in the bottom sediments they are exposed to, and are affected by, low but long-term doses of pollutants. They have been used to measure the degree of past pollution, in addition to indicating the current situation.

Vinegar fly
Drosophila melanogaster

This species of vinegar fly is found all over the world, but has found a particular home in biology laboratories! It is one of the most studied insects in the world because its short life cycle (a couple of weeks) makes it extremely useful in the study of genetics.

More commonly we know vinegar flies as nuisance flies that love fermenting or rotting fruit. As they are so small, normal fly-wire screens will not keep them outside. You will often find them near fruit bowls, if the fruit is a bit old, or in compost in very large numbers at times.

Vinegar flies have distinctive red eyes. However, genetic studies have identified many different mutants and strains. Some have different coloured eyes, no wings or other traits that would ensure they would not stand much of a chance of persisting in the wild.

Other species of *Drosophila* are also fairly common, and all look superficially alike. The immature insects are tiny maggots that develop quickly in rotting fruit. They are a problem at times in wineries, which of course have large vats of crushed fruit (grapes) fermenting to make wine, and also to producers of orange juice.

2 mm
Eats rotting fruit
Diptera

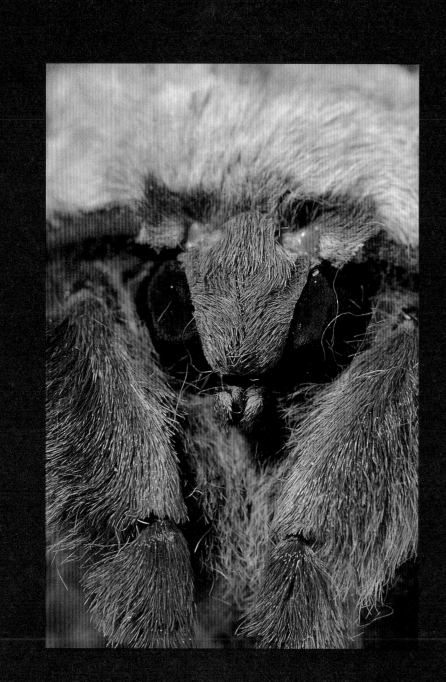

BUTTERFLIES, SKIPPERS AND MOTHS
Lepidoptera

What is the difference between a butterfly and a moth, and what is a skipper? It takes a closer look than you might imagine to distinguish between these three groups. The easiest way is to have a good look at the antennae ('feelers'); butterflies have a small club on the tip, skippers have a hooked tip, and moths generally have feather-like antennae. Generally butterflies are more colourful and active in the daytime while moths are duller coloured and active at night, but there are also common, brightly coloured, day-flying moths. Skippers look very much like butterflies but are generally faster flyers.

The colouring of these insects comes from a covering of tiny scales on the wings. These scales, and so the colour, come off easily if you touch them, leaving a dusty smear on your hands.

The winged butterflies, moths and skippers are the adult insects. The immature insects are caterpillars, which then undergo an intermediate pupal (or chrysalis) stage. Caterpillars have chewing mouth-parts and most eat leaves. The pupae do not feed, and the adults have (long) sucking mouth-parts (the proboscis) for different (liquid) foods such as nectar from flowers. There are species of butterflies known from South America that have adapted to feeding on another liquid: blood.

The ability of these insects to feed on different food sources in the different stages of their life cycle means that they can take advantage of the different types of food that are present only at certain times of year. That is, adults can fly to different areas to feed on flowers and nectar for the brief time they are present and then lay their eggs

where an appropriate food will be available for their caterpillar offspring. Some species are regarded as serious pests. Adults of such species as cabbage white butterfly (*Pieris rapae*), armyworm and budworm (both Noctuidae) find crops and then lay their eggs in this vast food source for the caterpillars, with few if any competitors. The caterpillars can develop rapidly and several generations can be completed during the life of the crop. The damage that results is sometimes devastating.

While some species are pests, most are not. Entomologists estimate that there are over 22 000 species of Lepidoptera in Australia, and only a tiny fraction of these are pests. Some of the pest species have been accidentally brought into Australia from overseas.

Common brown butterfly
Heteronympha merope

There are a few groups of butterflies that are easily recognized by their colouring; the 'browns', 'blues', 'whites' and 'coppers'. There are many species of 'browns' of which several commonly occur in urban areas. Females of the common brown are slightly larger than the male, and have yellow patches on the fore wings (as in the photograph) that the males lack. They have the small clubs on the tips of their antennae that are characteristic of butterflies.

The caterpillars of the common brown and other closely related species feed on grasses, usually climbing the stalks to do so at night and hiding under rocks or logs during the daytime. The adult female butterfly is known to lay her eggs directly on to several types of grass in order to provide the young caterpillars with a ready food source as soon as they hatch, or she may simply drop her eggs while flying over grass.

The adult butterflies emerge to feed on nectar in the warmer months of the year, taking advantage of a different food type from that available for the caterpillars. This means that they are active when both flowers and sunshine are present, in spring and summer.

60 mm (female, wingspan)
Eats plants
Lepidoptera

Orchard butterfly, citrus butterfly

Papilio * *aegeus*

The orchard butterfly is also known as the citrus butterfly because of the caterpillars' love for several types of citrus plants. Adult butterflies are attracted to flowers including many grown in suburban gardens.

The orchard butterfly is found from north Queensland (and Papua New Guinea) to as far south as some parts of Victoria and South Australia. Male and female orchard butterflies look very different from each other. The female is larger than the male, and has semi-transparent fore wings and brightly spotted hind wings. The photograph shows a male resting on a plant.

In addition to feeding on commercially grown citrus plants (such as oranges, lemons and grapefruit), orchard butterflies feed on native plants in the same family such as the native lime. Orchard butterflies can be minor pests in citrus orchards, but they are also attacked by many other insects (including wasp and fly parasites) which help to control their numbers.

The immature insects use different means of defence. Larger caterpillars protect themselves with some spines and also a pair of red 'tentacles'. These tentacles give off a smell that is supposed to ward off potential predators. Very young caterpillars have a different means of protection—they look like bird droppings.

100 mm (wingspan)
Eats plants
Lepidoptera

* Some scientists and books call this genus *Princeps*.

Australian painted lady
Vanessa kershawi

There are butterflies known as 'painted ladies' throughout the world. Australia has its own species which is extremely difficult to distinguish from the others. Indeed, until quite recently it was considered to be the same species.

The Australian painted lady migrates from place to place and can often be found in very large numbers. It can be seen in any month in northern Australia but in southern Australia it is essentially found from spring to autumn. It is surprising to see just how many of them appear after just a few warm, sunny days at the end of winter.

These butterflies spread their wings out horizontally when they stop to rest or to feed. They are attracted to many types of flowering plants and the caterpillars will also feed on a range of plants including native daisies and introduced capeweed.

43 mm (wingspan)
Eats plants
Lepidoptera

Wanderer
Danaus plexippus

As its name suggests, the wanderer butterfly roams great distances. It is found in many countries and in Australia it occurs in all states. Overseas it is also known as the monarch.

The Australian wanderer is not known to migrate, but in North America the monarch migrates south for the winter and then returns north for the summer. Wintering monarchs in America form large swarms in particular places and can literally cover trees. They are known to form wintering groups, although not in such numbers, near Sydney and Adelaide.

The caterpillars feed on a group of plants that have a smelly, milky sap. These caterpillars have bright bands across their bodies that warn potential predators (such as birds) that they taste horrible or are poisonous. It is likely that they use noxious compounds from their food plants to give them such protection. The pupal cocoon, in contrast, is a beautiful, decorated, blue-green container awaiting the emergence of the adult.

Male and female wanderers look very much alike, but the male has a dark spot in the middle of each hind wing (the main photograph shows a female). These spots are sacs that contain a scent believed to attract female wanderers.

40 mm (larva), 16 mm (cocoon), 95 mm (adult, wingspan)
Eats plants
Lepidoptera

Common eggfly
Hypolimnas bolina

Common throughout the tropical north at any time of year, the common eggfly may also (less often) be found as far south as Victoria and South Australia.

Males and females look quite different, but both have dark wings with incredibly iridescent patches of white and blue. The female is more spectacular than the male but is also quite variable. She usually has an orange patch lacking on the male. The photograph shows a male eggfly.

Adults are commonly seen feeding on flowers in home gardens in the north. The caterpillars are less commonly seen, as they feed at night on several different plant species. They protect themselves with an array of spikes along the body and head.

Common eggflies are not only found in Australia, but also occur in Papua New Guinea and Indonesia, and on the islands to the east of Australia.

80 mm (male, wingspan)
Eats plants
Lepidoptera

Cabbage white butterfly
Pieris rapae

Cabbage whites are very common at certain times of the year, especially around cabbages and other plants in the brassica group. Brassicas include plants often grown in home vegetable gardens such as cabbages, cauliflowers, brussels sprouts and broccoli.

The caterpillars are pests that eat these vegetables and so, although the butterfly is pleasant to look at, many people would prefer not to have them around their gardens. (In America the caterpillars are known as cabbageworm.) Cabbage whites are a pest in the northern hemisphere and are relative newcomers to Australia. They are believed to have arrived in the 1930s and have done so well that they are now found over much of this country.

When the caterpillars are ready to pupate, they move to safe places, such as fences or posts, or the underside of the food-plant. They attach themselves and remain there for perhaps only a few days until the adult butterfly emerges.

Male and female cabbage white butterflies have slightly different patterns on their wings. Females have two obvious black spots on each fore wing, but the males have only one. (Only the undersides of the hind wings can be seen in this photograph.)

43 mm (wingspan)
Eats plants
Lepidoptera

Birdwing butterfly
Troides* priamus

The birdwings are our largest Australian butterflies and, with the spectacular colours of the males, are bound to catch the attention of people seeing them sail past. They are not widespread compared to other butterflies, but they are possibly our most famous and striking species.

The colours on the wings of males are very different from those of females. Females are also generally larger than the males. They are usually found in rainforest in north-eastern Australia. Related species and subspecies occur as far south as northern New South Wales (the Cape York birdwing is shown in the photograph). Adult butterflies feed on flowers while the caterpillars feed on climbing plants in the genus *Aristolochia*.

For protection, the caterpillars have rows of large spines along their bodies. The eggs from which they develop are large (as might be expected) and laid singly, not in clusters. The large caterpillars are capable of eating all the leaves on their food-plants; if too many caterpillars were on a single plant none might survive.

There is quite a degree of variation in the colours and markings in the adult females. However, their size alone should make them fairly easy to recognize if one crosses your path!

120 mm (wingspan)
Eats plants
Lepidoptera

---

* This birdwing butterfly was until recently known as *Ornithoptera priamus*. (*Ornitho-ptera* literally means bird-wing.)

Common imperial blue butterfly
Jalmenus evagoras

The imperial blue butterfly has several interesting features. First is the fact that caterpillars and pupae are always surrounded, and so protected, by ants (as in the photograph). The ants eat the honeydew produced by the caterpillars and in return help to keep the imperial blue's insect enemies away.

Imperial blue butterflies are always found on wattle trees (*Acacia*), but the type of wattle varies from place to place. They are found in the mainland capital cities in eastern Australia (Melbourne just makes it into their known range). The number of generations per year varies with latitude. There are fewer in the cooler southern areas, more in the north.

Another interesting feature of the imperial blue is the 'tail' on the bottom of each hind wing. When the butterflies are sitting, these tails wave and look like antennae. It is easy to imagine that any predator may be fooled into thinking that the insect's head was under these 'antennae'. If the predator then lunged at the wrong end of the butterfly, chances are that the butterfly would escape—although perhaps with a damaged hind wing! The blue is only on the top side of the fore wing; the butterflies are mainly brown underneath. (You can see the blue colouring more clearly in the photograph opposite the title page.)

All stages of this species can occur together around a single wattle tree. It is a lovely sight to see these beautiful butterflies and the ant-attended caterpillars and pupae in summer.

20 mm (larva), 35 mm (adult, wingspan)
Eats plants
Lepidoptera

Common grass-blue butterfly
Zizina labradus

According to the experts (Common and Waterhouse, see 'Suggested Further Reading' section) this is the commonest butterfly in Australia. So you should have seen one of these!

The common grass-blue is most often found in open areas (such as the lawn in your garden) and in large numbers. You can encounter them in most areas of Australia throughout the year, but in the south they are more common in spring and summer.

The caterpillars feed on legumes (plants such as peas, beans and clover). There is a plentiful supply of food for them in most areas, and they have taken advantage of it.

The adults have a very delicate blue-purple colouration. When they are at rest, as in the photograph, the underside of the wings is shown, and that is a very pale blue-grey colour. It is only when the butterflies take flight that the darker flash of blue appears.

23 mm (wingspan)
Eats plants
Lepidoptera

Skipper
Hesperiidae

Skippers are similar to butterflies, distinguished mainly by the difference in their antennae. Butterflies have a club at the end of their antennae but skippers have more of a hook shape (clearly shown in this photograph of a species of *Toxidia*). Most skippers resemble small butterflies, but their flight is faster. This gives them their name, as they seem to 'skip' from point to point.

Skippers are active on warm, sunny days and feed from a variety of flowers. They are a common sight in gardens in spring and summer, as the flowers are very attractive to them. If you live on a hilltop then you are more likely to see gatherings of male skippers than your neighbours below, as they regularly gather in elevated areas and may even maintain a small territory.

Many species of skippers have very similar markings, making them difficult to tell apart. There are closely related species that look almost identical but are found in quite different areas.

The caterpillars are not as obvious as the adults, as they usually feed at night. Caterpillars of most species feed on grasses. An ideal site for skippers would be native grasslands (for the caterpillars) near to sunny, open, floral gardens by a hilltop.

27 mm (wingspan)
Eats plants
Lepidoptera

Swift moth
Hepialidae

These large moths—the one in the photograph belongs to the genus *Abantiades*—are often found at lights after rain in autumn and winter. Swift moths are interesting as they do not feed (and so do not live very long), hang by their fore legs, and the females can produce thousands of eggs each. To lay their eggs, some species simply drop them like bombs from an aeroplane as they fly over a suitable host plant.

The reddish-brown pupal skins are possibly more often found than the adults. These are left poking out of the ground where the adults emerged, often under red gum or grey box trees.

The larvae live in the soil feeding on the roots of these and other trees. Autumn rains appear to stimulate the pupae to emerge and often many moths will emerge at the same time. The odd sight of empty pupal cases protruding like mushrooms from the ground can happen just overnight.

Other species of swift moth will tunnel into the wood of trees, not simply feeding on the roots. Others again are grass feeders and a small number are considered pasture pests.

90 mm (wingspan)

Eats plant roots

Lepidoptera

Gumleaf skeletonizer
Uraba lugens

A notable caterpillar in many areas, especially along creeks or rivers with river red gums is the gumleaf skeletonizer, *Uraba lugens*. As its name suggests, this species feeds on eucalypt leaves until only the skeleton remains. Many types of eucalypts are attacked, but the red gums seem to be particular favourites.

The caterpillars are very hairy, and are often found in clusters on leaves. The hairs are for defence, and are the tips of poison glands on the body. When a potential enemy touches the hairs, the caterpillar releases a venom which can sting the attacker. Even the moulted skin of the caterpillar can still sting. When the caterpillar spins its cocoon before becoming a pupa, it incorporates some stinging hairs which help to protect it from enemies as it cannot move away.

Gumleaf skeletonizers have another unusual habit. Every time the caterpillars shed their skin, they keep the old scalp (head capsule) and glue it behind their head. After a few moults they have a fairly large 'hat', earning them the nickname of 'haterpillars'.

There may be two generations of caterpillars a year, one in summer and one in winter. Young caterpillars cluster together on the leaves of eucalypts but larger caterpillars like to have a leaf to themselves. Smaller caterpillars probably need to cluster together for protection, but larger individuals can protect themselves.

10 mm (larva),
20 mm (adult,
wingspan)
Eats leaves
Lepidoptera

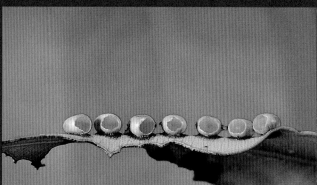

Emperor gum moth
Opodiphthera* eucalypti

The large emperor gum moth is well known in eastern Australian cities. Adult moths have prominent 'eye-spots' on each of their four large wings, designed to startle birds that might want to eat them. The moths lift the front pair of wings to suddenly expose another pair of brighter, larger 'eyes' looking back at the potential predator.

Adult females glue groups of eggs on to leaves, usually those of food-plants for the caterpillars. The caterpillars are well known to many, particularly school children who often find them and keep them as pets. The larger caterpillars feed voraciously on eucalypt leaves, and it is surprising just how many leaves need to be collected to keep them growing. The caterpillars have brightly coloured protective spines on their green bodies, and a yellowish stripe down their sides.

Caterpillars ready to turn into pupae spin a very tough silken cocoon, usually reinforcing it with wood or bark. The insect may 'rest' inside the cocoon to survive harsh conditions, and it may be one or more years before an adult moth emerges. They emerge in summer, but need to release a special liquid to soften the tough cocoon before it can be cut open. Apart from their eye spots, the moths can be recognized by their strong, curved fore wings, which often appear to be hooked at the end. Their colour varies; they can be brown, yellow or grey. (The photographs on pp. 134 and 136 show the head of an adult moth and a caterpillar.)

2–3 mm (egg),
65 mm (larva),
125 mm (adult, wingspan)
Eats plants
Lepidoptera

---

*An earlier scientific name of this species was *Antheraea eucalypti*.

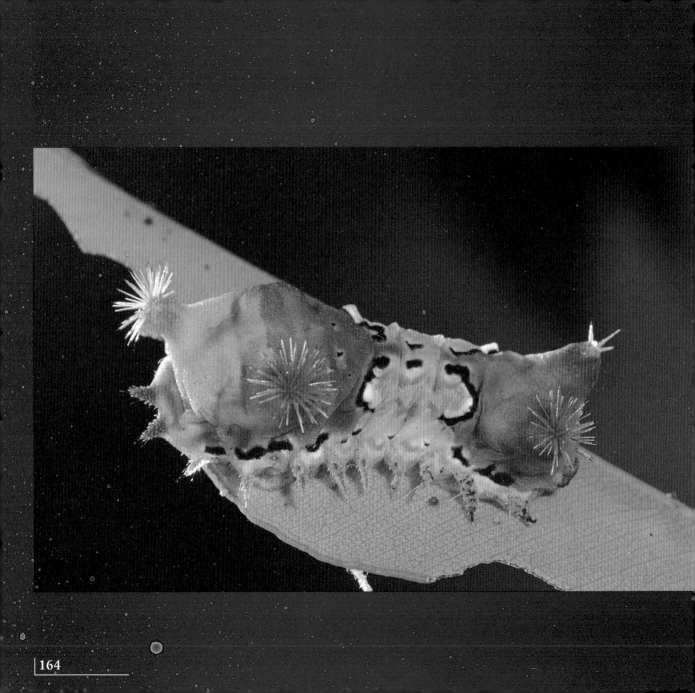

Cup moth
Limacodidae

The brightly coloured cup moth caterpillar and the pupa's distinctive, cup-shaped case make these insects stand out among the moths. They are commonly found feeding on eucalypts but will also attack fruit trees and other ornamental garden plants. In some cases they can be serious pests, defoliating trees.

The cup moths we see most often (and the caterpillar in the photograph) belong to the genus *Doratifera*, and there are several species with similar larvae. The spines on these caterpillars can give painful stings, so be careful if you want a close look at the beautiful colours. The stinging spines are found in clusters that the caterpillar retracts when they are not needed. When the caterpillar is disturbed, it pops out the rosettes of spines to sting whatever might want to eat (or collect) it. The spines will not actually puncture your skin, but the poison is released if the tips of spines are broken.

The caterpillars are sometimes also known by other names such as slug moths, Chinese junks (*Doratifera oxleyi*) or Bondi trams (*Doratifera vulnerans*). This is because they have shorter legs than most caterpillars, which makes them look as if they are gliding along rather than walking.

The name cup moth is given because of the hard pupal cocoon, the top of which is sliced off like the top of a boiled egg when the adult moth emerges. The lid or cap often remains attached by a hinge.

15 mm (larva), 45 mm (adult, wingspan)
Eats leaves
Lepidoptera

Case moth
Psychidae

The most interesting thing about a case moth is the cocoon made by the caterpillar. It is amazingly precise in its construction, and the covering can usually be used to identify the species. The photograph shows the cocoon made by the faggot case moth, *Clania ignobilis*. All the sticks are the same length except for one, and all individuals of this species construct exactly the same type of cocoon. The leaf case moth, *Hyalarcta huebneri*, constructs a cocoon out of the fragments of leaves of whatever tree it is feeding upon.

As the caterpillars crawl around with their cases attached to them, their legs and thorax are well developed but they have relatively small abdomens. When disturbed they withdraw into their cases and pull the opening shut, like drawing a shawl over their heads. The protection given by the case must be worth the effort of making and carrying it, because the adult females of some species as well as the larvae remain in them.

Clania ignobilis feeds on many eucalypts, but related species also feed on a wider range of native plants. Some case moths collect in large numbers on a single tree, but more often they are found alone.

30 mm (case)
Eats leaves
Lepidoptera

Grapevine moth
Phalaenoides glycinae

This species is one of an unusual few: a day-flying moth. It is like others in its sub-family group (Agaristinae) in having quite bright colours. In other words, it looks and acts in some ways more like a butterfly rather than a moth.

The grapevine moth is a native Australian species, usually feeding on native plants such as guinea-flower (*Hibbertia*). However, it has taken a liking to grapevines and can find even your isolated backyard plants as well as larger vineyards. The caterpillars grow fairly large and can cause quite a lot of damage to the leaves before they pupate.

The grapevine moth, with a characteristic orange tip to its abdomen, is found in all the eastern states and can be a pest in some vineyards in South Australia. The cater-pillars are also brightly coloured, suggesting that they must be distasteful or poison-ous to predators (such as birds). As they make no attempt to conceal themselves, the coloration is probably a warning not to eat them.

45 mm
(wingspan)
Eats plants
Lepidoptera

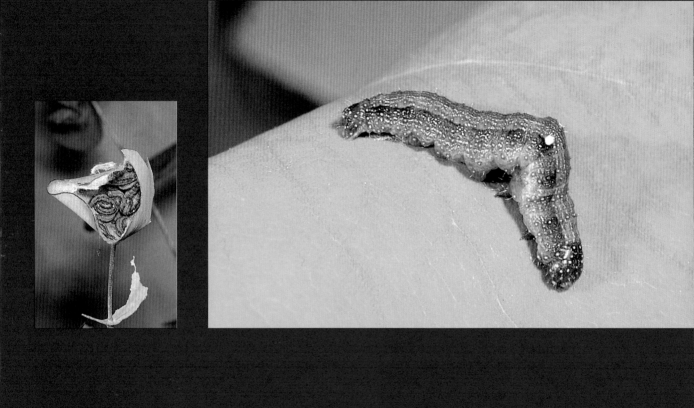

Autumn gum moth
Mnesampela privata

Caterpillars of this species are typical of many in its family (the Geometridae) and are called 'loopers'. The name comes from the way the caterpillar walks, moving the back part of its body so the middle arches up in a loop, and then moving the front half of its body. Autumn gum moth caterpillars increase the visual effect with a pair of bright yellow dots on their middles, marking the top of the loop.

30 mm (larva), 40 mm (adult, wingspan)
Eats leaves
Lepidoptera

Autumn gum moth caterpillars love to eat young eucalypt leaves, and so find young trees particularly attractive. When the young caterpillars hatch they eat only one side of the leaf, but larger, older caterpillars eat the entire leaf. They can kill young trees and are found from late summer to early spring. So watch for them on your developing, newly planted gums!

The caterpillars make themselves shelters in which to hide, and more than one caterpillar can be found in a shelter (as in the photograph). They construct them by curling over a leaf and holding it fast with silk threads. On severely damaged trees, it is only the leaves that form these shelters that remain! Usually the caterpillars are active at night, retreating into their shelters by day.

Caterpillars ready to pupate burrow into the soil near the tree they fed on, and the pupae shelter in the soil. The adults emerge in autumn and any female moth should be in just the right place to find a mate and lay her batches of eggs.

Codling moth
Cydia pomonella

Codling moths are pests of apples and are found in most parts of the world, although, so far, no permanent populations have been found in Western Australia.

The adult female moth lays her eggs on small, developing apples and the leaves of apple trees. Soon after the young caterpillar hatches it tunnels into the apple. There it feeds under the skin before heading for the seeds in the apple core. It grows into a large caterpillar within the apple and then chews its way to the surface.

The caterpillar leaves the apple and finds a sheltered place on the ground or low down on the apple tree to pupate. One way to help control this pest is based on this feature of the insect's biology. You can put a collar of corrugated cardboard around the trunk of a tree on which the mature caterpillars build their cocoons. You simply check regularly under the cardboard and kill any codling moth caterpillars or pupae.

There can be more than one generation of codling moths per year, depending on the temperature. Usually the species spends winter as caterpillars, pupating in spring with adults emerging in time for the first apples.

Not all caterpillars that eat your apples are codling moths. There is also a native of Australia, the lightbrown apple moth (*Epiphyas postvittana*), that has taken a liking to apples and grapes and is now also a pest overseas.

17 mm (wingspan)
Eats plants
Lepidoptera

Bogong moth
Agrotis infusa

Many farmers know these insects as cutworm and it is the caterpillars that worry them. The caterpillars eat pasture and crops and so are a pest. To most other people, it is the adult moth that is of importance, both as a pest and a useful insect.

Before white settlement, native Australians feasted on the adult moths in the Alps. The tribes knew the life cycle of the insects and that the adult moths (yet to mate) would spend the summer in the relative cool of the mountains. Rich in fat, because they were still to produce offspring, they were a good food-source.

Since white settlement, bogong moths have made themselves known to people living in the eastern capital cities of Melbourne, Canberra and Sydney. A 'plague' of bogong moths occurred in Sydney as long ago as 1867. In Canberra they recently invaded the new Parliament House, thinking no doubt that it was simply a convenient dark, quiet cave. They sheltered in wall cavities and similar dark places, where they became plentiful food for other animals such as mice and rats.

Bogong moths migrate long distances (several hundred kilometres) in late spring from lowland grassy areas to the summer alpine retreats. In autumn they make the return journey from the mountains to the plains. If the winds blow the wrong way, many end up dying at sea between Australia and New Zealand and are then washed up on Australian beaches.

There are two common species of *Agrotis*, both commonly called cutworm caterpillars. They are the common cutworm or bogong moth (*Agrotis infusa*) and the brown cutworm (*Agrotis munda*).

42 mm (wingspan)
Eats plants
Lepidoptera

Clothes moth
Tineidae

There are several different species of moths that we commonly call clothes moth, all belonging to the family Tineidae. The most common of all is an accidentally introduced pest, *Tineola bisselliella*. The immature caterpillars (like the one in the photograph) live sheltered within silken galleries, like cocoons, from which they emerge to feed on woollen cloth, jumpers, sheepskins, carpets, and similar materials. When they are found away from humans they would normally feed on things like feathers, fur, dead animals and, of course, wool.

The species in this family evolved to scavenge on dead animals or on animal products in places like nests, dens and lairs. The adult moths generally avoid light, in contrast to most moths which fly to lights at night. To them, a human home full of wool (carpet) and cloth in dark cupboards and wardrobes is just heaven!

In nature, clothes moths are an important part of the food chain, and play an important role in breaking down animal products. They are a key part of some cave communities where they feed on animal droppings.

10 mm (larva)
Scavenger
Lepidoptera

Magpie moth
Nyctemera amica

What colours would you expect on a magpie and what on a tiger? Well, this magpie moth is one of the tiger moth family (Arctiidae) and has both markings. Its wings are black with white blotches, and its abdomen has black and yellow bands.

The magpie moth is quite common in much of Australia, flies during the daytime and has brightly coloured markings. All these things together suggest that it must taste pretty terrible to predators! It is likely that these moths obtain a chemical defence as caterpillars by feeding on plants containing toxic chemicals.

Since the caterpillars obtain the noxious chemicals, they, as may be expected, are also brightly coloured to warn predators not to eat them. They look like the abdomen of the adult, with black and orange markings.

As well as occurring in Australia, magpie moths are found in countries to the north and in the Pacific islands. Adult moths are found in the summer (from late December) in the south of Australia.

40 mm (wingspan)
Eats plants
Lepidoptera

BEES, WASPS, ANTS AND SAWFLIES
Hymenoptera

THE HYMENOPTERA IS A LARGE, diverse group and, on an evolutionary time-scale, is one of the most advanced insect groups. It contains many of the social insects, which live in communities and have different castes responsible for different duties within the community.

The common honey bee (*Apis mellifera*) is perhaps the best-known social insect, with queens, workers and soldiers contributing to produce honey. In addition to the social species, there are also many Australian bees that are solitary species. Ants and many wasps are well-known social insects, but sawflies, although you can see them in groups, are not social insects.

A few wasps, such as the European wasp, are pests, but wasps in general are over-whelmingly beneficial insects. Parasitic species are often used in preference to predatory species in biological control programmes because the wasps are often highly selective of their victims (hosts). They will not change to feeding on non-pest species after they have eliminated the pest.

You may wonder what it is that makes bees, wasps, sawflies and ants different from all other insects, but similar to each other. Adults usually have a very narrow 'waist', and larvae are usually maggot-like, with no legs but with a distinct pair of jaws. To confuse the story, adult sawflies lack the waist and the larvae do have legs. Hymenoptera can be easily distinguished from flies (which only have one pair of wings) as they have two pairs of wings. The wings have a coupling mechanism to link the fore and hind wings in flight. Females often have an ovipositor, which, in many species, is also used for stinging. Ants are a distinct family, Formicidae, within the Hymenoptera.

Honey bee
Apis mellifera

Because it produces honey and pollinates plants, the common honey bee is well known and well thought of despite the fact that it stings. It is not native to Australia but was one of the first insects introduced to this country.

Honey bees have a very structured, complex society, with kings (called drones) and queens, workers, and nurseries. The type of food given to the larva determines what role it will have as an adult. Future kings and queens are provided with 'royal jelly' but the others miss out.

Worker bees visit flowers and collect pollen and nectar. They have special sacs on their legs to carry what they collect back to the hive. When they return, one of the most amazing communication processes in the animal world takes place. A returning bee tells the others how far away and in what direction it just collected food. It does this by a figure-eight 'waggle dance', which indicates the relative position of the sun and the plant, and also the richness of the food.

A strain of very gentle bees (Ligurian bees), which are not inclined to sting, was imported to Kangaroo Island many years ago. They remained pure bred while the European bees did not, so the Australian version is now in much demand.

Not everyone considers the introduced honey bee to be a good thing as it can compete for resources with native bees, and so reduce their numbers. Feral bees (not maintained in hives) are a particular problem, and can also harbour bee diseases.

14 mm
Eats pollen
Hymenoptera

European wasp
Vespula germanica

Unfortunately, one insect that has become all too common around southern Australian cities in the last twenty years is the distinctive European wasp. Often an unwelcome guest at barbecues or outdoor parties, the European wasp is a scavenger which also has a powerful sting. Black and yellow, it looks like a bee but, unlike bees, is able to sting repeatedly, not just once. You may notice them around the front grilles of parked cars, where they feed on the squashed insects stuck there. They are common in spring, summer and autumn. (The photograph on p. viii also shows a European wasp.)

The European wasp was accidentally introduced into Australia. It, and a very closely related species—the English or common wasp (*Vespula vulgaris*)—have quickly colonized suburban Melbourne and other cities. (European wasps are more numerous.) Both are social insects, living in colonies established by a queen. She is slightly larger than a worker, and may be seen searching for suitable crevices in wall cavities, tree hollows or underground in which to make an intricate nest. The wasps chew wood or bark or flowers to make a pulp and form small cells cemented together, to make a paper home.

The cream-coloured, legless, maggot-like, helpless larvae develop in the cells, fed by the workers until they pupate. The winged adults that emerge may be reproductive males or queens, or non-reproductive workers depending on environmental factors. In Europe the severe winters kill many of the wasps and prevent the continuous expansion of colonies. In Australia, with our comparatively mild winters, nests can grow very large and house thousands of wasps.

18 mm
Scavenger, predator
Hymenoptera

Mud-dauber wasp, sand wasp
Sphecidae

Not only do mud-dauber wasps provide their young with food, they also build them a mud-brick house! The adult female wasp first selects a suitable site, usually in a crevice or groove so that some of the sides already exist (as in the photograph), and then constructs a 'house' with bits of mud. Each bit of mud is laid like a brick upon the previous layer.

The outside of the nest may be rough, but the inside is smooth. In it, the adult female wasp places a paralysed host, usually a particular type of insect or a spider. (The species illustrated obviously prefers spiders, but is not too fussy about which types.) She then lays her egg on the host. The immature wasp develops over the next weeks as a maggot-like grub on the outside of the host (the arrow on the photograph points to such a grub). Some similar species in this family continue to provide fresh food, that is, new parasitized hosts, for their developing young. Other species are sealed into their cells with enough food to last until they emerge as adult wasps.

Nests are often constructed in a cluster or in a line. The source of mud for the nests may differ between years, leading to different coloured cells being produced.

The wasps' habit of selecting sheltered sites can cause some problems when they block machinery (for example, radiators, exhaust pipes and even handlebars on push-bikes). They have also caused problems where they have chosen the same rocks as those used by Aborigines for cave paintings and the nests destroy the artwork.

4 mm (larva),
18 mm (adult)
Parasite
Hymenoptera

Wasp
Netelia sp.

This is a large, but slender, reddish-orange wasp that is attracted to lights at night. It is a parasite of caterpillars but it can also sting people!

    The adult wasps feed on nectar and dew drops, not on caterpillars. The adult female wasp lays an egg on a caterpillar and the egg then hatches into a parasitic maggot. Caterpillars with parasites look as though they have a pale, tiny bean attached to them. It is actually the developing larva (maggot) of the wasp.

20 mm
Parasite
Hymenoptera

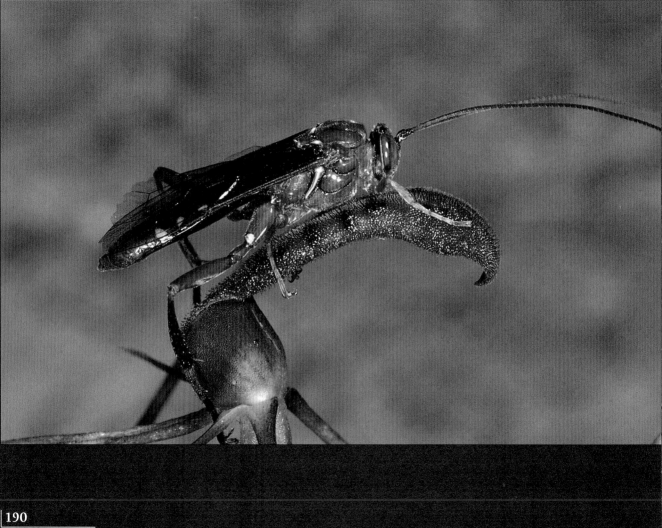

Orchid dupe
Lissopimpla sp.

This attractively coloured wasp is commonly found in the daytime searching for hosts. It attacks caterpillars, usually large individuals about to pupate, or the pupae themselves. The large egg-laying ovipositor appears very sturdy, perhaps because the adult female wasp has to push it through stiff tussock grasses where the hosts are hiding.

There are orchids that are successful in tricking male *Lissopimpla* into trying to mate with them. The short-sighted male wasps believe the flowers are female wasps and, in their efforts to attempt mating, can help to pollinate the flower.

11 mm
Parasite
Hymenoptera

Gall-formers

Not only are many species of Hymenoptera gall-formers, but there are also many flies (Diptera), true bugs (Hemiptera) and beetles (Coleoptera) which cause plants to form galls or large swellings. We have chosen this gall from an acacia as an example of this habit.

Gall-formers cause plants to form galls so that their young are provided with food and shelter. However, galls are often little communities of several species. While one species may have caused the gall, others often invade. The invaders may feed on the gall tissue, feed on the gall-former or become a parasite of other insects in the gall.

If you break open a young, developing gall (still green) you will find little maggot-like grubs inside. Old, woody galls are usually empty, with neat round holes all over them (like those on the left of the bottom photograph). The holes indicate that the insects have completed their development and the adults have emerged.

Galls on eucalypts are often formed by the combined action of a fly and a nematode (a roundworm). Both benefit from the gall and so they co-operate in its construction. Not all galls are caused by insects; many are also caused by fungi. Galls on oak trees are commonly called oak apples.

1 mm (larva)
Eats plants
Hymenoptera

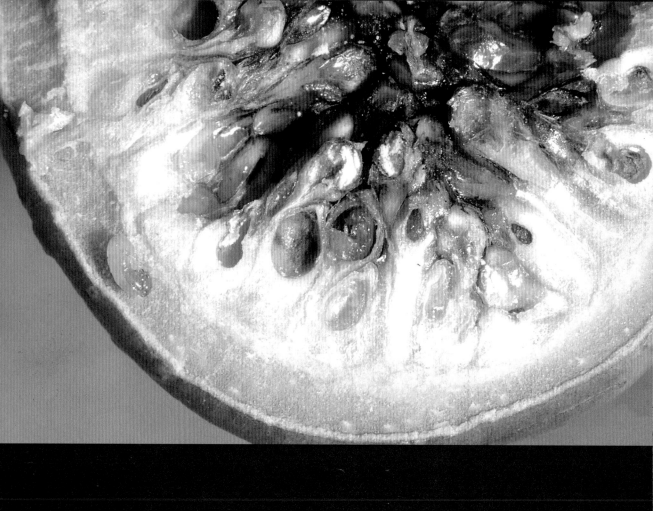

Fig wasps
Agaonidae

These amazing wasps all live within fruits produced by various species of fig (*Ficus*). The relationship is of benefit, in fact essential, to both the fig and the wasp. The wasp gains food and shelter and the fig is pollinated as its flowers are on the *inside* of the fruit!

Female wasps begin the process by either inserting their eggs through the walls of the figs or by squeezing completely inside the fig to lay eggs. Developing larvae feed on the tissue within the fig, some forming galls. You can clearly see the larva in the photograph, on the fig's left edge.

Adult male fig wasps are often wingless and they may never see outside the fig but females are winged and do leave the fruit. (So perhaps there is more protein in a fig than you had imagined!) The adults pollinate the fig. Wasps are very particular about the species of figs they live in and each wasp will only develop in one species of fig. However, some species of fig will contain more than one wasp species.

You may think that the discovery of such a minute, secretive life was something fairly recent in biological science. However, that is not so. A Roman called Pliny who lived in the first century AD described the close relationship between fig wasps and cultivated figs!

1.5 mm
Eats plants
Hymenoptera

Bull ant, bulldog ant*
Myrmecia

The painful sting of a bull ant is something that most of us remember! Their nests are seemingly irresistible to children, who feel challenged to poke sticks down an entrance. However, the cunning ants often use an alternative exit to attack their molesters from behind.

Our common bull ants are unique to Australia. They are one of the world's largest ants and are called bulldog ants because of their fierce grip once they bite. The jaws (seen more clearly in the photograph on p. 180) are only used to hold on firmly while the ant jabs its sting, found at the tip of its abdomen, deep into its victim.

Bull ants, like most other ants, have different castes. The queen is winged until she forms a colony. Males are always winged, but soldiers and workers are wingless. All live in a colony but, unlike wasps and bees, bull ants do not make individual brood chambers. The young live in a communal chamber.

Bull ants are independent, solitary hunters and scavengers. Unlike so many other ants they do not form trails; instead, individuals seek out food and carry it back to their nest.

Jumping jacks, which jump as well as sting, also belong to the genus *Myrmecia*. Another *Myrmecia* species is very unusual in the animal world as it has only one pair of chromosomes. (Humans have twenty-three pairs.)

23 mm
Predator, scavenger
Hymenoptera: Formicidae

---

* While many entomology books call *Myrmecia* 'bulldog ants', most people call them 'bull ants'. There are other ants with very big heads but no sting that, at least in days gone by, were also called bull ants.

Meat ant
Iridomyrmex purpureus

Their distinctive mounds make meat ants most obvious. These ants carefully maintain a covering of gravel-sized objects on the roof of their nests (as in the photograph). This clearly marks where they live and probably helps to stop the roof from cracking open.

During winter (in southern areas at least) meat ants are not very active. But at other times they will, if disturbed, surge from their nests to combat intruders. They detect the arrival of something (or someone) from the vibrations made by the movements. If it is an approaching enemy they will attack it; if it is prey, they will kill the animal and take it back to the nest. Meat ants do not sting like bull ants or jumping ants but still give a good bite. When scores of them bite at the same time they will chase away most visitors!

Trails from the nest are usually clearly visible in vegetation. These are formed by the marching feet (six at a time of course) of thousands of ants going to and from the nest each day. These trails are the equivalent of our freeways, but instead of white lines they use a chemical to mark the way.

A large central nest can have satellite colonies. New colonies begin when a fertilized queen lands in unclaimed territory or when a satellite becomes independent. A queen ant may rule for fifteen years or more over an increasing colony with complex underground galleries. When the colony is well established, she may rule hundreds of thousands of worker ants.

8 mm
Predator, scavenger
Hymenoptera: Formicidae

Argentine ant
*Linepithema humile**

These common urban pests are found in many countries and, as the name suggests, were first found in Argentina. They are nuisance pests, invading our homes but not biting, stinging or causing damage. Like some other ants, they can protect scale insects and other honeydew-producing species. They can then help make these pests a worse problem. In their defence, it has been suggested that Argentine ants are sometimes beneficial because they attack termites.

Argentine ants live in fairly shallow nests, often under stones. When we put down paving, concrete blocks or asphalt paths we provide them with an ideal habitat. That is one reason that the ants flourish in urban areas.

Intensive chemical eradication programmes were attempted in some cities, notably Perth and Sydney. However, after many years without success, these programmes were abandoned. Argentine ants can form huge colonies, with budding, or the splitting off, of colonies common. The separate colonies seem to live together peacefully, but together they will attack other (native) ant species. They are known to compete with, and replace, many native ants and as such cause even further disruption to the environment.

2 mm
Generalist feeder
Hymenoptera: Formicidae

---

* Argentine ants had the name *Iridomyrmex humilis* until 1992, and so you will find them under that name in most books.

Sugar ant
Camponotus sp.

Orange-and-black coloration is usually a warning sign, but these ants are not the aggressive, stinging sort. Although they are large, like bull ants, they are different in many ways.

Commonly called sugar ants, they live in large colonies and workers head out on marked trails to forage. The ants leave chemical markers for others to follow so that, if they find a good, large food supply, they can then co-operate to bring it back to the nest. Sugar ants can be seen during the daytime, but they are more active at night.

At certain times, colonies produce winged adults who disperse to colonize new areas. These adults gather near the entrance to the nest so that when the conditions are right, they can pour out together.

Although they do not sting, the larger soldiers will still give you a nip. They have much larger heads than the workers and the main reason for that is to accommodate the muscles to work their powerful jaws.

10 mm (worker), 13 mm (soldier)

Generalist feeder

Hymenoptera: Formicidae

Green tree ant
Oecophylla smaragdina

Most people in tropical Australia will be familiar with the green tree ant, and its bite! It is not pleasant if you interrupt a trail of these ants with a bare arm or leg. They move very quickly and not only bite but squirt a dose of acid into the wound to make doubly sure you know they are angry.

Green tree ants do not nest in the ground but, as their name suggests, they make leafy nests in trees. The way they weave their nests takes a great deal of co-operation between individual ants, and also requires the help of their young. This is unusual, even for social insects.

The nests are constructed by sticking leaves together with silk. The adult ants cannot produce the silk; it is made by the larvae. So, while some workers bring together the two leaves that will be joined, others carry silk-producing larvae and touch the larvae first on one leaf, then the other. The larvae are really used like a needle and thread combined to stitch the leaves to each other. As nests can be made from more than just two leaves, the process is repeated until a complete shelter is constructed.

9 mm
Scavenger
Hymenoptera:
Formicidae

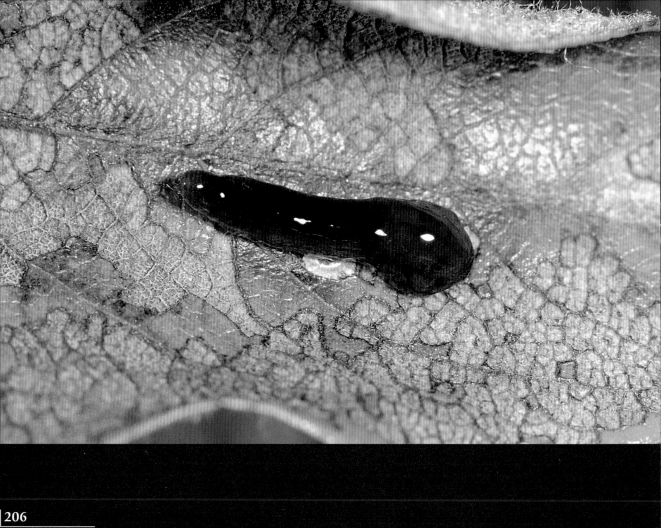

Pear and cherry slug

Caliroa cerasi

One of the most unusual insects commonly seen in gardens is a type of sawfly. At first glance it does not seem like an insect at all, but looks like a small, flat slug. It selects pear, cherry, plum, quince and hawthorn trees and some ornamental plant leaves to feed on, stripping them back to the bare leaf veins. The adult is a winged, black, glossy insect closely related to wasps and resembling a small fly. It is the young sawflies that look like slugs.

There are two generations a year in southern Australia with the slugs appearing in late spring and again in late summer. The adult emerges in spring from its pupal cocoon in the soil, beneath the trees it attacks, and where the female lays her eggs. She cuts slits in the leaves with her saw-like, egg-laying structure (ovipositor) and deposits her eggs inside.

A single tree can have a large number of pear and cherry slugs attacking it and the damage can be serious. This insect is not native to Australia but is an introduced pest that almost certainly arrived with the trees it feeds on. (Its preferred foods are also not native to Australia.)

The slug-like larvae can range in colour from dark-green (almost black) to orange. They cover themselves with slime and have no obvious legs or other appendages when looked at from above.

10 mm
Eats leaves
Hymenoptera

Sawfly
Perga sp.

Often mistaken for repulsive caterpillars, these insects are sawfly larvae and are closely related to bees and wasps. They actually try to look as repulsive as possible.

The larvae cluster together in the daytime and, if disturbed, rear up and vomit a blob of yellowish liquid. This liquid is a concentrate of eucalyptus oils obtained from the leaves they feed on, which they store in a special sac until needed. You see a revolting, Medusa-like mass of seething snakes, a sight generally enough to put off anything that might want to eat them. This behaviour has earned them the name 'spitfires'.

The larvae eat eucalypt leaves, and usually start with the youngest leaves on the tips of branches. So damage is usually noticed at the top of a tree first (as leafless branches appear) and then lower down as the insects move down the tree.

Sawflies are actually very clever, as the larvae can communicate with each other. At night, when they spread out to feed on a tree, they keep in touch by using an equivalent of Morse code. They tap on the branches with the tips of their abdomens so that the other members of the group know where they are and so that they can re-form the cluster before daylight. The larvae feel the vibrations rather than hear any sound.

Adult females have a serrated, saw-like structure (an ovipositor) with which to cut a slit in suitable trees for their eggs. That is where the name 'sawfly' originates.

40 mm (larva),
25 mm (adult)
Eats leaves
Hymenoptera

NON-INSECTS

Spiders
Araneida

Spiders are most easily distinguished from insects by having eight legs (not six) and only two (not three) body sections, the cephalothorax (which includes the head) and the abdomen. They have a distinct waist separating these two sections. They do not have chewing mouth-parts, but instead kill their prey with poison, dissolve the edible parts with enzymes and, finally, suck up the soup! (The spider in the photograph measured approximately 60 mm from toe to toe.)

Scorpions
Scorpionida

Scorpions are predators, related to spiders, but they have a long tail with a sting at its tip. Their front legs have large claws (like a crab's) which are used to hold their prey. They use the sting to subdue their prey before eating it. This scorpion is 30 mm long.

Slaters
Isopoda

Known variously as woodlice, pillbugs and butcher-boys, slaters are members of the Crustaceae. That is, they are part of a group more often found in water than on land, as the Crustaceae includes crabs, crayfish and shrimps. Slaters are flattened animals that live in moist places and, if disturbed, roll themselves into a ball like an armadillo. These slaters are 10 mm long.

Slugs
Gastropoda

Slugs (and snails) are invertebrates but not arthropods. They move by sliding on their 'stomach-foot' which they lubricate to ease their progress. Unfortunately, most of the slugs and snails we see in our gardens are species that have been accidentally introduced from overseas. (The species in the photograph is the great yellow slug, *Lehmannia flava*, and it is 110 mm long.) Native species tend to avoid cultivated gardens but some can still be found in native bush gardens.

Millipedes and Centipedes
Diplopoda and Chilopoda

Millipedes and centipedes are somewhat similar-looking animals, one difference being that millipedes have two pairs of legs per body segment while centipedes have only one pair. Their names reflect this difference as millipede means 1000 legs and centipede means 100 legs. The Portuguese millipede (shown in the upper photograph) is a relatively new arrival to Australia, and is a significant nuisance pest in some cities such as Adelaide and Melbourne. Centipedes are predatory, feeding on a wide range of other invertebrates. This Portuguese millipede is 50 mm long, while the centipede measures 90 mm in length.

APPENDIX 1
COLLECTING INSECTS

I F YOU ARE HAPPY TO WATCH insects as they arrive in your garden, or wherever you are, then you don't really need to make a collection. If, on the other hand, you want to keep some to look at or study whenever you want, then making an insect collection is worth while. We will give you an outline of what is involved, but if you are serious about making a collection then we suggest you read more detailed texts (see p. 225) or join an entomological society and learn from an experienced collector.

The main tools of an insect collector are containers and a net. The net is useful for capturing flying insects but many specimens can be easily caught without one. For example, insects attracted to lights and windows at night and insects under logs, rocks or on bushes are all fairly easy to collect. (Just watch out when turning logs and rocks, as other animals such as spiders and snakes also live in these places.)

Once caught, the insects should be killed and then preserved. They can be killed with chemicals (such as ethyl-acetate) or simply put in the freezer. When killed, but before going brittle or mouldy (usually a couple of days), the insects should be mounted on special entomological pins. The wings, body and legs are arranged ('set') as desired and then left to dry. This may take a month or more, depending on the size of the insect.

When dried, mounted insects should be kept in an airtight box. These are specially made for insect collecting, with foam or cork lining to hold the pins. Naphthalene or moth balls (held in place with pins or gauze) should be put in the box to repel insects that may eat your specimens. The whole box should be kept in a dry place.

Equipment and books on collecting insects can be obtained from specialist suppliers. One is Australian Entomological Supplies, PO Box 250, Bangalow, NSW 2479.

APPENDIX 2
PHOTOGRAPHING INSECTS

INSECT PHOTOGRAPHY IS DIFFERENT from other types of nature photography. The small size of most insects makes it necessary to use special techniques and equipment but the results can be very rewarding (and quite spectacular) if you can follow a few simple rules. In case you were wondering, all the insects shown in this book were photographed while alive.

Equipment. What do you need?

Insect photography falls into the category of macrophotography or, more precisely, photomacrography. The most essential item, of course, is the *camera*. For this type of macro work, Denis recommends a reliable 35 mm SLR camera with interchangeable lenses. The camera body need not be particularly sophisticated, but some form of through-the-lens (TTL) exposure metering will be useful.

The main requirement for the *lenses* is that they are able to focus closely on an insect so that the subject fills the frame. A normal (50 mm) lens with some form of extension tube between the camera body and lens can do this. All the tubes do is allow the lens to focus more closely, and the magnification gained depends on the length of the extension tubes. That means that you would use longer extension tubes to photograph smaller insects.

Even better than extension tubes is a true *macro* lens. This type of lens is better because it is designed to give flat fields, corner-to-corner sharpness, and to function better in the macro ranges than conventional lenses of the same focal length. A magnification of 1:1 (that is, the photograph comes out the same size as the subject being photographed) is very useful. Most modern macro lenses can reach a magnification of 1:1 without extension tubes. That means that a subject that is 36 mm long will fill the frame on a 35 mm camera. A macro lens with a focal length of 100 mm is the most useful of all. This lens allows a greater working distance between the lens and subject than a 50 mm lens, which makes it less likely that the subject will be disturbed while the photograph is being taken.

A sturdy *tripod* is essential if natural light is being used and it is also advisable when you are using a flash, particularly at magnifications greater than 1:1. The depth of field in a frame is shortened as magnification increases and so you will find it more difficult to obtain sharp images. The slightest movement of the camera or subject will result in a blurred photograph. You will need a set of *bellows* to gain magnifications greater than 1:1. An example would be if you wanted to fill the frame (36 mm) with a single ladybird beetle (5 mm).

Denis uses *flash* to light most subjects because the results are generally sharper and more predictable than those in natural light. A single flash unit is often enough to give natural-looking results. You may find that harsh shadows can be a problem, but you can soften these by taping a piece of tracing paper over the flash head. You could also tape a piece of white card to the lens on the opposite side to the flash. What happens then is that the flash lights the subject and throws a shadow, but some light will hit the card and bounce back into the shadow side of the subject. You will only learn how to position the card correctly by experimenting.

Flash-lit insect photographs often have black backgrounds, because the flash is lighting the subject but not the background. You can avoid this by either waiting for the insect to move into a better position or by moving something like a branch in behind the subject. Denis usually photographs insects indoors which gives a greater degree of control over the final results. Insects can move very quickly, but, being 'cold-blooded', they can also be slowed down by placing them in a refrigerator before your photographic session. This normally does not harm them but merely immobilizes them.

You should use *film* of the finest grain possible, as it will give sharper results. The lower the ISO number of a film, the sharper it is. An ISO 50 film would be ideal—but keep in mind that lower-speed films require more light.

Exposure

Some cameras have automatic exposure systems for natural light and flash. If you have a manual camera then you should take a meter reading and determine a correction because, as magnification increases, less light reaches the film due to the extension between lens and film.

To determine the correct magnification, and then the correction you require for proper exposure, you can use the following procedure. *Magnification* (m) is determined by dividing the *image size* by the *object size*. The easiest way to do this is to place a ruler across the field of view and read the distance through the viewfinder. For example, if the field of view is 72 mm on the ruler (object size) and as the width of a frame of 35 mm film is 36 mm (image size), simply divide 36 by 72 and the result is 0.5× or 1:2. Magnifications can be read off the barrel of the lens, but they are often inaccurate and are only useful up to 1:1.

$$\textit{Exposure correction} = (m + 1)^2$$

Using our example, exposure correction = $(0.5 + 1)^2 = 2.25$. This means that we need twice as much exposure, which simply means an extra *f* stop in camera lens terms.

These calculations take a bit of getting used to, but they are necessary unless the camera has an automatic exposure system.

If natural light is being used, the camera's in-built light meter should be sufficient. A flash meter should be used if flashes are being employed, but the $(m+1)^2$ rule can be used in conjunction with your flash's guide number calculations (see your flash instruction manual). Experimentation and bracketing are essential before you will feel comfortable with this set of rules. Some books that will help you with your photographs of insects are listed below.

Suggested reading

Cooper, J. D. and Abbot, J. C. *Nikon Handbook Series: Close-up Photography and Copying*. Amphoto, New York, 1979.

Imes, R. *The Practical Entomologist*. Aurum Press, London, 1992.

Shaw, J. *John Shaw's Close-ups in Nature*. Phaidon Press, Oxford, 1987.

White, W. *Close-up Photography and Photomacrography*. Eastman Kodak, Rochester, New York, 1984.

GLOSSARY

abdomen The parts of an insect behind the legs.

antennae The 'feelers'.

caste A particular form of a social insect (that is, either a worker, soldier, or reproductive).

caterpillar Larval stage of a moth, skipper or butterfly.

cerci 'tails' on the tip of the abdomen.

chrysalis Pupal stage of an insect, where a larva turns into an adult.

cuticle The insect skin.

detritus Usually mulch or rotting vegetation.

diapause A dormant or resting stage, usually for the pupa.

elytra The hardened fore wings of a beetle, modified to form a cover over the abdomen.

entomologist A person who studies insects.

entomology The study of insects.

exoskeleton The hard, jointed skin of an insect that forms the equivalent of a skeleton.

family A grouping of similar genera and species within an order.

forceps Opposable appendages that can be used to grasp objects.

generalist feeder An insect that will eat a wide range of food, including plants and animals.

haemocoele The blood-like fluid that fills the body cavity.

halteres The hind wings of a fly, reduced to small stubby clubs.

head capsule The hard cuticle that forms the skeleton of the insect head. It is moulted and shed as the insect develops.

invertebrate An animal without a backbone.

larva The immature stage of an insect.

nymph The immature stage of an insect that does not have a pupal stage.

ootheca An egg-case.

order A grouping of families that are similar. There are twenty-nine orders of insects.

ovipositor Tube-like body-part in some female insects used for laying eggs.

parasite An animal (or other organism) that develops in, or on, another living animal or plant.

predator An animal that kills and eats other animals.

proboscis Mouth-parts, developed into an extendible tube.

pronotum A distinct plate on the back or upper side of the first segment of the thorax in beetles.

pupa The immobile stage of an insect's development where a larva turns into an adult.

pupate When a larva changes to a pupa.

siphon A tube used like a snorkel by aquatic or semi-aquatic insects to breathe air.

social Insects with divisions of labour (i.e. workers, soldiers, queens).

thorax The three body segments behind the head that carry the wings and the legs of an insect.

wing-bud A developing wing on an immature insect.

SUGGESTED FURTHER READING

Common, I. B. F. *Moths of Australia*. Melbourne University Press, Carlton, 1990.

Common, I. B. F. and Waterhouse, D. F. *Butterflies of Australia*. Angus & Robertson, Sydney, 1982.

CSIRO (ed.). *The Insects of Australia* (2nd edn). Melbourne University Press, Carlton, 1991.

CSIRO. *Insects—A World of Diversity* (CD-ROM). CSIRO, Melbourne, 1994.

Dahms, E. C., Monteith, G. and Monteith, S. *Collecting, Preserving and Classifying Insects*. Queensland Museum, Brisbane, 1990.

Harvey, M. S. and Yen, A. L. *Worms to Wasps*. Oxford University Press, Melbourne, 1989

Hawkeswood, T. J. *Beetles of Australia*. Angus & Robertson, Sydney, 1987.

Hughes, R. D. *Living Insects*. Collins, Sydney, 1974.

Naumann, I. D. *CSIRO Handbook of Australian Insect Names*. (6th edn) CSIRO, Melbourne, 1993.

New, T. R. *A Guide to Insects of South-eastern Australia*. Oxford University Press, Melbourne (in press).

New, T. R. *Insects as Predators*. New South Wales University Press, Kensington, New South Wales, 1991.

New, T. R. *Introductory Entomology for Australian Students*. New South Wales University Press, Kensington, New South Wales, 1992.

Zborowski, P. and Storey, R. *A Field Guide to Insects in Australia*. Reed, Chatswood, New South Wales, 1995.

INDEX